Innovative Strategies for Solid Waste Minimisation in the Mining and Minerals Industry

Innovative Strategies for Solid Waste Minimisation in the Mining and Minerals Industry

Editors

Chongchong Qi
Guichen Li
Hakan Basarir
Qiusong Chen
Yuantian Sun

MDPI

Basel • Beijing • Wuhan • Barcelona • Belgrade • Novi Sad • Cluj • Manchester

Editors
Chongchong Qi
School of Resources and Safety Engineering
Central South University
Changsha
China

Guichen Li
School of Mines
China University of Mining and Technology
Xuzhou
China

Hakan Basarir
Department of Geoscience and Petroleum
Norwegian University of Science and Technology
Trondheim
Norway

Qiusong Chen
School of Resources and Safety Engineering
Central South University
Changsha
China

Yuantian Sun
School of Mines
China University of Mining and Technology
Xuzhou
China

Editorial Office
MDPI AG
Grosspeteranlage 5
4052 Basel, Switzerland

This is a reprint of articles from the Special Issue published online in the open access journal *Minerals* (ISSN 2075-163X) (available at: https://www.mdpi.com/journal/minerals/special_issues/ISSWMMMI).

For citation purposes, cite each article independently as indicated on the article page online and as indicated below:

Lastname, A.A.; Lastname, B.B. Article Title. *Journal Name* **Year**, *Volume Number*, Page Range.

ISBN 978-3-7258-2357-4 (Hbk)
ISBN 978-3-7258-2358-1 (PDF)
doi.org/10.3390/books978-3-7258-2358-1

Contents

minerals

MDPI

Article

Waste Slag from Heating Plants as a Partial Replacement for Cement in Mortar and Concrete Production. Part I—Physical–Chemical and Physical–Mechanical Characterization of Slag

Andrijana Nedeljković [1], Marija Stojmenović [1], Jelena Gulicovski [1], Nenad Ristić [2], Sonja Milićević [3], Jugoslav Krstić [4] and Milan Kragović [1,*]

1 "Vinča" Institute of Nuclear Sciences, National Institute of the Republic of Serbia, University of Belgrade, 22-24 Mike Petrovića Alasa, 11351 Belgrade, Serbia; andrijana.nedeljkovic@vinca.rs (A.N.); mpusevac@vinca.rs (M.S.); rocenj@vinca.rs (J.G.)

2 Faculty of Civil Engineering and Architecture, University of Niš, 18000 Niš, Serbia; nenad.ristic@gaf.ni.ac.rs

3 Institute for Technology of Nuclear and Other Mineral Raw Materials, Franše d'Eperea 86, 11000 Belgrade, Serbia; s.milicevic@itnms.ac.rs

4 University of Belgrade, Institute of Chemistry, Technology and Metallurgy, National Institute, Centre for Catalysis and Chemical Engineering, 11000 Belgrade, Serbia.; jkrstic@nanosys.ihtm.bg.ac.rs

* Correspondence: m.kragovic@vinca.rs

Received: 10 October 2020; Accepted: 6 November 2020; Published: 9 November 2020

Abstract: Numerous factors influence the complexity of environmental and waste management problems, and the most significant goal is the reuse of materials that have completed their "life cycle" and the reduction in the use of new resources. In order to reduce impact of waste slag on the environment, in the present study, waste slag, generated in heating plants after lignite combustion, was characterized in detail and tested for application as a replacement for cement in mortar or concrete production. For physical–chemical characterization of slag, different experimental and instrumental techniques were used such as chemical composition and determination of the content of heavy metals, investigation of morphological and textural properties, thermal analysis, X-ray, and infrared spectroscopy. Physical–mechanical characterization of slag was also performed and included determination of activity index, water requirement, setting time and soundness. A leaching test was also performed. Presented results show that waste slag may be used in mortar and concrete production as a partial cement replacement, but after additional combustion at 650 °C and partial replacement of slag with silica fume in the minimal amount of 12%. The maximal obtained cement replacement was 20% (17.8% slag and 2.2% of silica fume).

Keywords: waste slag; lignite; cement; environmental protection; mortar and concrete

1. Introduction

Coal is the energy source with the highest percentage of estimated fossil fuels (over 65%), with more evenly distributed deposits in the world than oil and gas. The USA, Russia, Australia, China, India, Indonesia, Germany, Ukraine, Poland, and Kazakhstan are some of the largest coal producers in the world [1]. Coal is a widespread fossil fuel used around the world. About 23% of the primary energy demand and 39% of electricity is obtained from coal. Due to the decline in oil and gas reserves, significant use of coal is expected to increase in the future. Today, about 40% of the world's total energy is produced from coal, and it is expected that this situation will remain in the future [2].

In general, there are various methods for the classification of coal, including the carbon content, origin, purpose, age, thermal power, and other properties. Thus, according to the carbon content, there are the following types of coal: (1) anthracite, or "hard coal", contains carbon in the range of 86–97%. It is characterized by a brittle texture and lustrous sheen and due to the high heat value it is mainly used in industrial settings and the metal industry; (2) bituminous coal, often called "soft coal", possesses a slightly lower content of carbon in comparison to anthracite (45–86%) and is mostly used for electricity and steel production; (3) sub-bituminous coal has a carbon content of 35–45% and its primary field of application is electricity generation; (4) lignite ("brown coal") has a carbon content of 25–35%. Its characteristics are high moisture content, which sometimes reaches 66%, and crumbly texture, and it is often used for the generation of electricity [3]. It is also known as medium-calorie coal and due to its application and quantity, it represents an exceptional potential in energy terms. Although it contains a relatively high amount of moisture, the thermal power of lignite is relatively high—ranging from 10 to 20 MJ/kg. The combustible part of lignite consists of solid organic matter and volatiles, while moisture and ash are the non-combustible part. The non-combustible part, in addition to fly ash, which represents about 85–95% of the total amount of ash, contains slag in the amount of about 10% [4]. If we take into account that the world's annual consumption of lignite is more than 1600 million tons [5], it practically means that the production of fly ash in the world is more than hundreds of millions tons [5,6], while the calculated amount of slag is more than 160 million tons. In Serbia, the annual production of fly ash is around six million tons, and slag 500–600 thousand tons [7].

From a chemistry standpoint, coal ash (or shorter ash) is defined as a solid non-combustible residue that remains after coal annealing. Thus, several types of ash can be distinguished as follows: slag (bottom ash), the primary non-combustible residue extracted from the bottom of the boiler; boiler ash, larger classes of which are extracted from the boiler together with the flue gases and fly ash, the finest class, separated from the boiler with flue gases [8].

Ash and slag obtained as waste materials by incomplete combustion of coal such as lignite in power or heating plants usually contain different, very toxic, microelements such as arsenic, cadmium, chromium, copper, manganese, nickel, lead, selenium, zinc as well as their compounds, which are very harmful for the environment, natural resources and hence may affect the life and health of humans [9–11]. According to the EU's rule book "List of Waste" [12] as well as Serbia's "Rules on the manner of handling waste that have the characteristics of hazardous substance" [13], waste material that contains heavy metals and other hazardous substances must be classified as hazardous waste.

In most countries, including Serbia, there are appropriate regulations that necessitate stricter controls and additional measures for securing, disposing and transporting generated hazardous waste as well as additional financial costs for the companies to pay the state in the view of additional taxes [14]. However, despite financial costs, the most significant problem is the impact of such waste on nature and the environment. Waste material from landfills can relatively easily pollute the environment by scattering both air and soil. as well as water systems and resources and thereby directly pollute and endanger the health of animals and plants, but the most significant concern is the health and food of humans. To reduce the harmful impacts on the environment, as well as to avoid financial costs for the companies, it is of particular importance to find useful value in such a waste. In this way, the benefits may be multiple—environmental and financial—so all aspects of green chemistry and circular economy would be satisfied.

The potential application of fly ash in industries is fairly well described in the literature [15–20]. However, practical applications of slag are still not well described and sufficiently examined. The slag particles have large dimensions, and because of this they may be used as an aggregate, in road construction and in the brick industry, as well as in the production of so-called slag-blocks. Additionally, in the EU member states, slag is intensively and increasingly used to encapsulate landfills [21,22]. Different authors have investigated the application of the blast furnace slag, which is generated in industries as a partial cement replacement. Adu-Amankwah et al. [23] investigated the influence of limestone on the hydration of ternary slag cements. In their study, mixes with different amounts of cement–slag/quartz–limestone/quartz–anhydrite were prepared and tested. The amount of slag used was expressed as slag/quartz content, which was in the range of ~28–47%. In the study of the Sum et al. [24], optimization of gypsum and slag content in blended cement containing slag was investigated. For these experiments, blast furnace slag was also used. In this study, different amounts of gypsum (0–40%), slag (60–100%) and clinker (0–40%) were mixed with Portland cement. Authors concluded that a high cement strength in the ternary slag–gypsum–clinker system may be obtained for 10% of gypsum, 20–40% of clinker and 60–75% of slag as well as for 10–25% of gypsum, less than 5% of clinker and 70–85% of slag. In the study of the Yoon et al. [25], blast furnace slag-blended Portland cement was used and tested for cobalt (Co) immobilization. In this study, authors replaced 0–70% of the cement with blast furnace slag. Their results show that regarding the retention of cobalt ions (Co^{2+}), retention increases with increasing the slag content in the mixes and the best results were obtained for cement replacement of 70%. However, the amount of slag that is obtained and already presented in the landfills is significantly higher than actual needs, and therefore there is a need for new research studies that could enable or provide guidelines for the use of slag in other industries. It is important to emphasize that the slag obtained from different sources (blast furnaces, hitting plants, etc.) possess different chemical compositions and thus potential application for all of them must be checked, investigated, and proven. From the aspect of practical application, it is very important that the use of slag has no negative effects on the environment, only positive effects, which are reflected in the protection of natural aggregates, reduction of CO_2 emissions, energy savings and, above all, reduction in waste material.

Due to its rough particles (sizes between sand and gravel), which have a glass structure and make it convenient to be used to replace natural aggregates (such as sand, gravel, crushed granite, etc.), potential practical application of slag in the cement industry was investigated in this study. The starting waste slag was characterized in detail and solutions were proposed to eliminate the defects of the slag from the aspect of application in the cement industry, and finally, the physical–mechanical characterizations of the obtained slag–Portland cement mixtures were made.

2. Materials and Methods

2.1. Sampling

The waste slag was sampled from landfill of a heating plant located in Valjevo, Serbia. The slag was obtained during combustion of lignite. The samples were taken from six different places at the landfill from three different depths, 30 samples in total. After sampling, the samples were dried at 105 °C for 2 h and homogenized. Homogenized slag samples were crushed, ground and sieved so the size of the grain and the particles were 100% below 63 μm before analysis.

Portland cement (CEM I 42.5R) was bought from the company CRH, Popovac, Serbia. Silica fume was bought from the company "Sika".

2.2. Characterization

Determination of the chemical composition and heavy metal content in slag was obtained with classical chemical analysis [26], using the atomic absorption spectrophotometer A Analyst 300 (Perkin Elmer, Waltham, MA, USA) and the spectrophotometer Specol 1300 (Analytic Jena, Wegberg, Germany), using standard and internally developed methods.

Simultaneous non-isothermal thermogravimetric analysis (TGA) and differential thermal analysis (DTA) were used for testing the thermal properties of the slag and the Portland cement. For measuring, a Setaram Setsys Evolution 1750 instrument (Setaram, Lion, France) was used. Measuring was performed in the temperature range of 25 to 1000 °C with a heating rate of 10 min^{-1} in an air atmosphere, with a flow of 16 cm^3 min^{-1}. The mass of the sample was 15 mg.

The Sorptomatic 1990 Thermo Finnigan device (Thermo Fisher, Waltham, MA, USA) was used to measure the textural properties of the slag. The nitrogen adsorption–desorption isotherms were obtained by adsorption at −196 °C. Firstly, the slag was degassed at room temperature for 1 h and then, under the same residual pressure, for 16 h at 110 °C. For analyzing the isotherms, Software ADP Version 5.13 Thermo Electron (Thermo Fisher, Waltham, MA, USA) was used. The Brunauer–Emmett–Teller (BET) method was used for determination of the specific surface area (S_{BET}) of the slag [27]. The micropore volume (V_{mic}-DR) was determined by application of the Dubinin–Radushkevich (DR) equation on the nitrogen adsorption isotherms [28], while the Dollimore and Heal (DH) method [29] was used for measuring the mesopore volume (V_{meso}).

The structural properties and the composition of the crystal phase of the slag and Portland cement were determined by Roentgen X-ray diffraction analysis (XRD). For XRD analysis, a PHILIPS PW-1710 X-ray diffractometer (Philips, Amsterdam, Netherlands) with a scintillation counter and a curved graphite monochromator was used. The measuring was performed at room temperature at 0.02°/s, in the range of 4–65° 2θ. A voltage of 40 kV and a current of 30 mA was used, while 1° and 0.1 mm slits were used for the direction of the primary and diffracted beams.

The field emission scanning electron microscopy (FESEM) and the TESCAN Mira3 XMU instrument at 20 kV (Tescan, Brno–Kohoutovice, Czech Republic) were used for investigation of the morphological and microstructural properties of the slag. Prior to analysis, the slag samples were pre-coated with several nanometers of a thick gold layer, using the sputter coater Polaron SC503 Fision Instrument (Quorum Technologies Ltd., Lewes, UK). The chemical composition of the surfaces were determined by energy-dispersive X-ray spectroscopy (EDS) analysis in conjunction with an Oxford Instruments INCA X-sight system (Oxford Instruments, UK) attached to the Jeol JSM-6460 LV SEM (JEOL Ltd., Tokyo, Japan).

The Thermo Scientific "Nicolet iS50" instrument ((Thermo Fisher, Waltham, MA, USA) was used for infrared spectroscopic analyses (FTIR) of the slag and Portland cement. Investigations were performed in the transmission mode. In the background, the KBr spectrum was used. The wavelength range was 400–4000 cm^{-1}, while the resolution and number of scans were 2 cm^{-1} and 64, respectively. Two corrections—base line automatic correction and suppression of the atmosphere—were used after recording the spectra.

2.3. *Testing of the Slag's Pozzolanic Activity*

Reactions that occur between the active ingredients of pozzolan, such as lime and water, represent pozzolanic activity. The progress of the pozzolanic reaction is usually monitored by a decrease in the free lime in the system or an increase in the amounts of silica and aluminum, which are soluble in acid[105] by using the Florentin method. Additionally, pozzolanic activity can be monitored indirectly through the measurement of the activity index (AI). According to this method, a part of Portland cement is replaced with the appropriate pozzolan in a precisely defined amount, and then the properties of such mixtures are compared with the properties of a mixture that is 100% Portland cement, i.e., the activity index (AI) represents the ratio of the strength of the Portland cement mixture with and without pozzolan [30].

The pozzolanic activity includes the maximum amount of lime that pozzolan can encompass and the degree of reaction over the time or the reaction rate between a pozzolan and calcium ions (Ca^{2+}) or calcium-hydroxide ($Ca(OH)_2$) in the presence of water. The rate of the reaction is dependent on the specific surface area, chemical composition, and the active phase content in the pozzolan. Both factors depend on the nature, quality, and quantity of the active phases in the pozzolans [30].

In this study, the pozzolanic activity of slag was followed by measuring the activity index (AI). The measuring was in accordance with the International EN 450-1 standard. The AI was determined as the ratio of the compressive strength of the standard mortar prism, which was prepared with 75 wt.% of cement and 25 wt.% of slag, to the compressive strength of the standard mortar prism prepared with 100% cement, tested at the same age. The EN 450-1b standard defines that the AI of the mortar prism prepared with 75 wt.% of cement and 25 wt.% of the slag, after 28 and 90 days, must not be less than 75% and 85%, respectively, in comparison with the AI values obtained for the prism prepared with 100% cement. In addition to the activity index, the following tests were also carried out on binder material: water requirement (WR) according to EN 450-1 Annex B, setting time (ST, initial (IST) and final (FST)) according to EN 196-3 and soundness (S) according to EN 196-3.

3. Results

3.1. Chemical Analysis and Determination of the Content of Heavy Metals

Chemical composition is very important, and this is one of the main conditions that determines whether a material can be used and whether it finds practical application in the cement industry. For waste materials, which are the subject of research in this paper, necessary criteria prescribed by the EN 450-1:2012 standard must be satisfied in order to apply the material in the cement industry and concrete or mortar production. For this reason, chemical analyses of the waste slag have been performed. The results are shown in Table 1, where the results of the chemical composition of the Portland cement (CEM I 42.5R) and the values prescribed by the standard are also presented.

Table 1. Compositions (in weight percentages, wt.%) of the slag and CEM I 42.5R and values prescribed by the standard.

Parameter	SiO_2	Fe_2O_3	Al_2O_3	CaO	MgO	SO_3	Na_2O	K_2O	P_2O_5	TiO_2	LOI *
Slag	21.20	10.81	15.70	6.32	2.50	1.78	2.67	0.63	0.036	0.62	37.45
CEM I 42.5R	21.62	2.60	7.00	60.16	2.34	2.55	0.33	0.66	-		2.68
Standard	$70 < SiO_2 + Fe_2O_3 + Al_2O_3$			≤10.0	≤4.0	≤3.0	≤5.0	-	≤5.0		≤11.0

* LOI—loss of ignition. Annealing temperature 1050 °C; ignition time 2 h.

The comparison of the slag properties and the requirements prescribed by the standard shows that there are two criteria that waste slag, in starting form, does not satisfy. At first, the sum of the masses of SiO_2, Fe_2O_3 and Al_2O_3 (47.71%) in the slag is less than 70%. Then, the loss of ignition (LOI) of the slag (37.21%) is more than the 11% that is prescribed by the standard.

Since the loss of ignition of slag mainly comes from the loss of unburned organic constituents or physically adsorbed water during annealing, this disadvantage may be surpassed by annealing the slag at the temperature of 650 °C.

This temperature was determined by TGA/DTA analyses (Section 3.2), which confirmed that 650 °C is the necessary temperature for the loss of unburned organic constituents or physically adsorbed water. After the additional thermal treatment of the slag, the percentual chemical composition changed, so that all requirements prescribed by the standard were satisfied. Results are shown in Table 2.

Table 2. Compositions of the slag after additional thermal treatment (in weight percentages, wt.%).

Parameter	SiO_2	Fe_2O_3	Al_2O_3	CaO	MgO	SO_3	Na_2O	K_2O	TiO_2	P_2O_5
Slag	33.75	17.46	25.96	10.00	3.98	2.84	4.25	1.00	0.70	0.0573

In addition to the determination of the chemical composition, the content of heavy metals in slag was also measured and results are shown in Table 3.

Table 3. The content of heavy metals in slag. Concentration is expressed as mg/kg.

Sample	Cr	Cu	Li	Mn	Ni	Pb	Zn	Cd	Sn	Sb	Ag
Slag	138	51	31	600	110	480	36	2.2	<5	160	5.4
Limit value *	70				40	50				5.0	

* Lower limit value for the classification of waste as hazardous waste, according to the rulebook on the categories, testing and classification of waste, Official Journal of the Republic of Serbia No 56/2010.

According to the rulebook on the categories, testing and classification of waste—Official Journal of the Republic of Serbia No 56/2010—the content of the Cr, Ni, Pb and Sb exceed the maximum values allowed for the classification of waste as non-hazardous, so slag should be categorized as hazardous waste due to its characteristics. For the use of waste materials in the cement industry in Europe, including Serbia, the rules prescribed by the "Rulebook on restrictions and prohibitions on the production, placing on the market and use of chemicals" as well as the "Rulebook on the categories, testing and classification of waste" are in use. These rulebooks do not restrict the concentration of heavy metals in the waste material. The only limitation according to the rulebooks is the restriction of releasing any heavy metals from the end products (mortars or hardened mortars) during leaching. Thus, a leaching test was performed on the mortar mixture where 17.8% of CEM I 42.5R was replaced with slag (maximal allowed amount of slag that may be used—results are shown in Section 3.7). The testing procedure consists of the leaching of crushed hardened mortar (grain size smaller than 4 mm) with water, the liquid–solid ratio was 10 dm^3/kg, and the determination of the heavy metal concentrations was obtained from eluates. The leaching is performed by mixing the mixture for 24 h. For the test, the mortar prisms used for testing AI were crushed. The leaching test was carried out in accordance with the EN 12457-2 standard and the results are shown in Table 4:

Table 4. Concentration of the released elements after the leaching test. Concentration is expressed in $\mu g/dm^3$.

Released Element	Concentration	Allowed Values *
Cr	75	4000–12,000
Cu	81	6000–12,000
Mn	114	-
Ni	154	3000–7500
Pb	563	5000–15,000
Zn	668	6000–20,000
Cd	29	50–200
Ag	21	-

* According to the rulebook on permitted quantities of hazardous and harmful materials in agricultural land and water for irrigation and methods for testing—applicable in Serbia.

3.2. *Thermogravimetric and Differential Thermal Analysis (TGA/DTA)*

The thermal properties of the slag and Portland cement were investigated by the simultaneous non-isothermal TGA/DTA analysis and the results are shown in the Supplemental Materials (Figure S1a (for slag), S1b (for Portland cement)) and Table 5.

Table 5. Weight loss in different temperature intervals for slag and CEM I42.5R samples.

	Weight Loss, %				
Temperature	25–150 °C	200–650 °C	650–800 °C	800–1000 °C	25–1000 °C
Slag	4.00	32.78	0.61	0.001	38.31
Temperature	25–400 °C	400–1000 °C			25–1000 °C
CEM I 42.5.	0.65	6.87			7.52

In the DTA diagram of the slag (S1a), one endothermic and two exothermic peaks were observed. The endothermic peak of DTA curves was detected at 95 °C, originating from the dehydration of physically adsorbed water. Due to this process, weight loss of about 4% in the first temperature interval (25–214 °C) was observed. In the second temperature interval (214–600 °C), two exothermic peaks were observed at 468 and 542 °C. For many burning products, such as slags or ashes, the loss of ignition (LOI) content can be equated to the amount of unburned organic constituents. The presence of organic carbon in a sample may be manifested by the appearance of an exotherm, or exotherms in the range of 400–600 °C. If the DTA curve contains two exotherms, the lower temperature exotherm represents amorphous (isotropic) carbon, and the higher temperature is due to the crystalline (anisotropic) variety. Thus, these exotherms at 468 and 542 °C correspond to the reaction of the production of CO_2 and CO due to combustion of the organic constituents, which contain isotropic and anisotropic carbon, respectively [31]. The mass loss during these reactions, which occurs in the temperature interval 214–650 °C, may be read from the thermogravimetric curve and was about 33% and corresponds to the content of organics in the sample. From the aspect of the application of slag in the cement industry, it is very important to notice that the content of organics and unburned carbon is restricted due to: (1) the tendency to adsorb the air entrainment admixtures added to the concrete to prevent crack formation and propagation and (2) increasing water requirements and decreasing strength development in concrete [32,33].

As it may also be seen from Figure S1b, for CEM I 42.5R two characteristic peaks are visible on the DTA curves. The endothermic peak of the DTA curves was detected at 113 °C, originating from endothermic reaction dehydration i.e., removal of physically adsorbed water. This process was followed with weight loss of about 0.65% in the temperature range 25–400 °C. In the second temperature interval (400–1000 °C), an endothermic peak with high intensity was observed at 730 °C, which was accompanied with weight loss of 6.87% on the TGA curve. This peak and weight loss can be explained by decomposition of the CEM I 42.5R aluminate phase, i.e., portlandite decomposition [34–36], dehydroxylation of the calcium aluminate and the alumina hydrates and the decomposition of the calcite (carbonates) [34–37]. However, due to the low amount of carbonate mineral calcite in the Portland cement, this process was not visible on the DTA curve.

3.3. Textural Properties

The nitrogen adsorption–desorption isotherm of the slag sample is shown in Figure S2, while the calculated textural parameters are shown in Table 6.

Table 6. Properties of the slag sample.

Sample	S_{BET}, m^2/g	V_{meso}, cm^3/g	V_{mic}, cm^3/g
Slag	25.0	0.020	0.010

The position of the hysteresis loop ending is characteristic for measurements with nitrogen at −196 °C and cannot be related to the textural properties of the material, but can be related to the use of nitrogen (N_2) as an adsorbate (so-called tensile strength effect) [38,39]. Although there is a certain content of mesopores in the analyzed material, which is indicated by the presence of a hysteresis loop, the small volume of the mesopores (0.020 cm^3/g), the absence of a horizontal plate at high p/p_0 and the position of the hysteresis loop end exclude the possibility of categorizing slag as an exclusively mesoporous material. Therefore, the resulting isotherm can be classified as type II, which is typical for non-porous or macroporous materials (International Union of Pure and Applied Chemistry—IUPAC nomenclature) [38,39].

The standard "EN 197-1-Cement—Part 1 Composition, specifications and conformity criteria for common cements", which is applicable in the Republic of Serbia and the EU, prescribes that the ingredients for cement should have a specific surface area (S_{BET}), determined in agreement with ISO

9277, of a minimum of 15 m^2/g. The S_{BET} of the slag sample was determined and the calculated value was $S_{BET} = 25.0\ m^2/g$, which means that the slag sample meets the required criteria.

3.4. X-ray Structural Analysis (XRD)

For the determination of the crystal phase composition of the slag, XRD analysis was used and the results are shown in Figure 1. Additionally, XRD analysis of the CEM I 42.5R was performed and results are also shown in Figure 1.

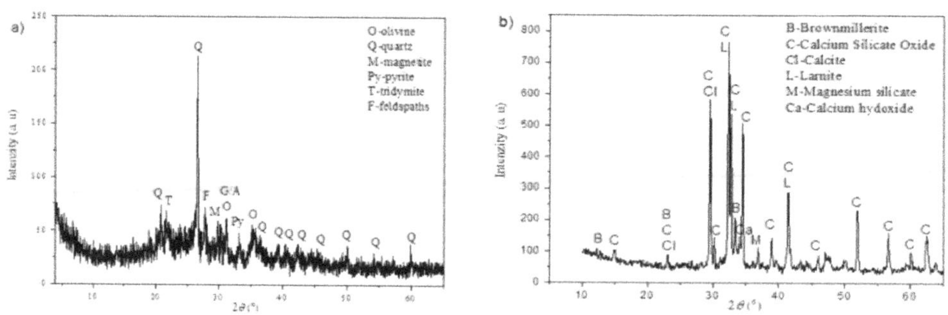

Figure 1. Results of the XRD analyses of the (**a**) slag and (**b**) CEM I 42.5R.

It was determined that the slag sample possesses low crystallinity. However, crystal phases were detected, and the following minerals/phases were present: quartz, olivine, magnetite, pyrite, tridymite, feldspars. The most abundant mineral is quartz, while olivines and feldspars and then all other minerals/phases are less present. Of the feldspars, plagioclases are dominant, relative to K-feldspars.

The results of the XRD analysis of the CEM I 42.5R sample show its high crystallinity and the following minerals were detected: brownmillerite, calcium silicate oxide, calcite, larnite, magnesium silicate, calcium hydroxide. The most abundant phase is calcium silicate oxide, then brownmillerite, larnite and calcite, while magnesium silicate and calcium hydroxide are present as associate minerals. Of the larnite, larnite-beta ($Ca_2(SiO_4)$) is dominant. The CEM I 42.5R crystallinity was high.

3.5. Scanning Electron Microscopy (SEM) Analysis

SEM images of the waste slag taken with different magnifications are shown at Figure 2.

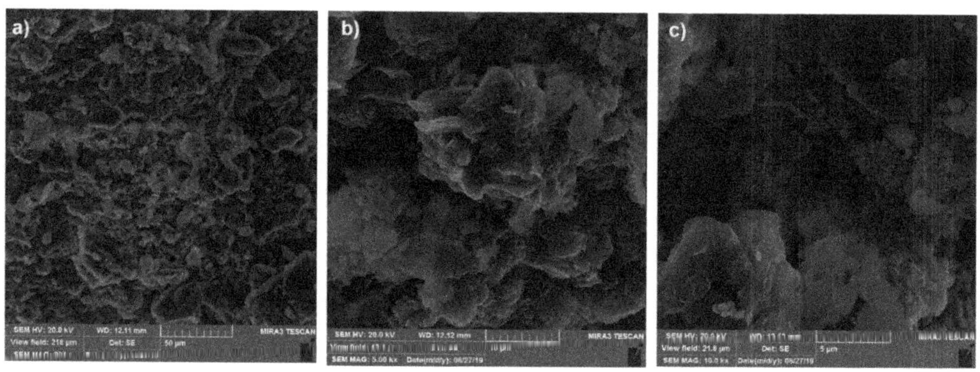

Figure 2. SEM micrographs of slag measured with different magnifications: (**a**) 994×, (**b**) 5000× and (**c**) 10000×.

From the micrographs taken with different magnifications, it can be seen that the slag is mainly constituted of aggregates of irregular size, which are mostly flat or have a granular structure.

Micrographs taken with higher magnifications (5000× and 10000×) show that slag is mainly constituted of irregularly shaped aggregates of different sizes in the range of 2 to 10 µm. From the morphology of the investigated sample, it is also evident that the slag sample possesses cavities, which most likely originate from the loss of the bound aggregates on the surface during the combustion process of lignite or the drying of the sample, such as water molecules from the surface, organic molecules or carbonates.

The chemical composition of the surfaces of the slag is presented in Figure 3 and Table 7.

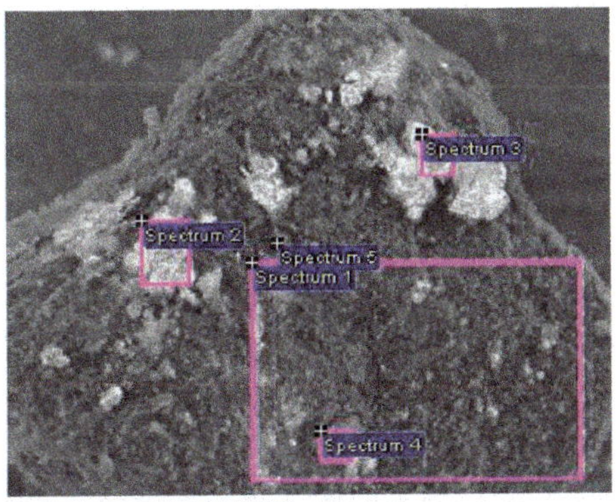

Figure 3. EDS-SEM micrography of waste slag.

Table 7. Composition of the slag surfaces at selected positions expressed in weight percentages (wt.%).

Spectrum Number	Element Concentration										
	C	O	Al	Si	Fe	Ti	Ca	Mg	K	Na	S
1	38.57	33.15	6.1	11.34	5.03	0.31	3.43	0.70	0.70	-	0.67
2	27.96	42.24	7.83	14.86	2.98	0.32	1.85	0.69	0.64	0.13	0.52
3	41.92	35.25	5.43	8.1	3.08	0.25	3.44	0.9	0.5	-	1.14
4	34.13	36.25	7.12	13.02	4,00	1.75	2.07	0.65	0.68	-	0.33
5	12.62	35.52	13.22	24.99	6.4	0.32	3.38	1.16	1.59	-	0.8

Five different positions and two measurement mods (spot and area) were used in order to get precise information about the chemical composition of the slag surface. Any chosen position for the EDS analysis of the surface composition of slag represents an aggregate of particles. This fact, pointed out previously in the analysis of the higher magnification micrographs, is confirmed by the analysis of the EDS spectrum results. The composition of the two spectra (one and five), of which spectrum one contains the results of the average composition of the surface at almost 0.1 mm^2, and spectrum five contains the results of the composition at a point in the immediate vicinity of the edge of said surface, differ significantly. Indeed, the carbon content at position five is three times smaller, while the contents of Aluminum and Silicon are over two times higher in relation to the average composition of the chosen surface (spectrum one). Taking into account the processes, which simultaneously occurred during the burning of lignite (dehydration, partially or complete oxidation) and slag cooling and settling (possible carbonation and rehydration) as well as the different phases identified by the XRD analysis, complex highly heterogeneous surface composition is expected. Nevertheless, a reasonable degree of agreement is obtained by comparing the average values of the major analyte of the EDS analysis with the results obtained by classical chemical analysis (Section 3.1).

3.6. *Infrared Spectroscopy with Fourier Transform (FTIR)*

The infrared spectra of the starting slag sample and CEM I 42.5R are shown in Figure 4. In the infrared spectrum of the slag, spectral bands at 3697, 3619, 3340, 1618, 1585, 1509, 1454, 1264, 1025,

1005, 911, 796, 777, 692, 522 and 460 cm^{-1} are presented. The spectral bands at 3697 and 3619 cm^{-1} are assigned to the stretching vibrations of structural hydroxyl groups in Al-OH and Si-OH [40,41]. The spectral bands at 3340 and 1618 cm^{-1} may be assigned to the presence of the physical bounded and surface adsorbed water and stretching and banding vibrations of OH^- groups, respectively [40,41]. The position of the spectral band is related to the banding vibrations of OH^- and groups of water molecules may be an indication that water molecules interact through oxygen atoms with surface cations and act as an H-bond acceptor. Namely, according to the literature, a spectral band is usually obtained at 1630 cm^{-1} and shifting in its position may give information about the acidity/alkalinity of the surface. When a water molecule acts as an H-bond acceptor, the δ (H_2O) is shifted to lower wave numbers; conversely, when it acts as H-bond donor, the δ (H_2O) is shifted to higher wave numbers [42]. Spectral bands at 1025, 1005, 796, 777, 692, 520 and 460 cm^{-1} can be assigned to the Si-O vibrations. According to their positions and shape, they are clear confirmation of the presence of quartz, plagioclases and tridymite in the waste slag, which were also confirmed by XRD analysis [43–45]. Additional confirmation of the presence of the plagioclases (albite ($NaAlSi_3O_8$) and anorthite ($CaAl_2Si_2O_8$)) is the spectral band at 911 cm^{-1}, which originated from Al-Al-OH banding vibrations [40]. Bands originating from other minerals presented in the crystal phase of the slag sample were not visible due to them overlapping with much more pronounced bands in the spectrum. Spectral bands at 1585, 1509, 1454 and 1264 cm^{-1} originated from slag components that are in the amorphous phase and were not visible by XRD analysis. These spectral bands may be assigned to the organic components of the slag. Lignite, whose incomplete combustion produces slag, which is the subject of this research, is humus coal from deposits in Serbia, Kolubara basin and belongs to type III kerogen, which is formed from terrestrial plants, rich in cellulose and lignin, with low lipid content [46,47]. Such coals are mainly aromatic in nature, so the vibrations visible in the spectrum of the slag sample are mainly related to the vibrations in the aromatic rings. Thus, spectral bands at 1509 and 1454 cm^{-1} are assigned to the ν_{19a} and ν_{19b} in-plane, aromatic ring vibrations. Vibrations ν_{8a} and ν_{8b} (in-plane, aromatic ring stretch) are assigned in the FTIR spectrum with spectral bands at 1603 and 1585 cm^{-1}. However, due to the overlapping of the band with lower intensity at 1603 cm^{-1} with those that originated from the vibration of the –OH group (at 1618 cm^{-1}), only the line at 1585 cm^{-1} is visible in the spectrum. Stronger intensity of the band at 1585 cm^{-1} in comparison to the band at 1603 cm^{-1} indicates that the substituent on the ring is conjugated to the ring [48]. Finally, the spectral band at 1264 cm^{-1} may be assigned to enolic stretching C-O vibrations [49].

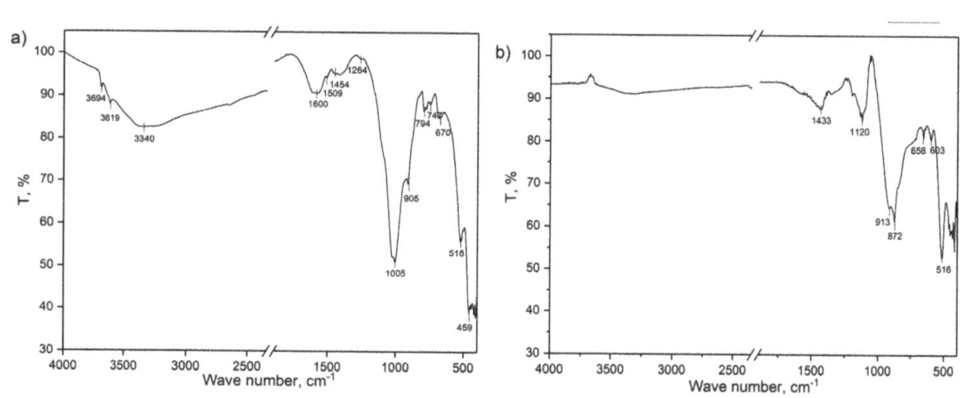

Figure 4. Infrared spectra of the: (**a**) slag and (**b**) CEM I 42.5R.

In the spectrum of the CEM I 42.5R, the band of 1433cm^{-1} is attributed to stretching vibrations of the CO_3^{-2}, and originates from the presence of calcium-carbonate ($CaCO_3$) in cement [8,41] and also, this spectral band is characteristic for brownmillerite [50], which was also detected by the XRD analysis. The spectral band at 914 cm^{-1} originated from Al-Al-OH bending vibrations, and the band at

876 cm^{-1} may be assigned to Fe-Al-OH deformations, which also confirmed the presence of brown millerite ($Ca_2(AlFe)_2O_5$) while the band at 516 cm^{-1} originated from Al-O-Si bending vibrations. The band at 657 cm $^{-1}$ is due to coupled Al-O and Si-O out of plane vibrations [8,39], while the band at 602 cm^{-1} is due to vibrations of internal modes of FeO_6 in brown millerite. The bands at 420 and 465 cm^{-1} originated from O-Mg-O and O-Ca-O bending modes in minerals, which contain calcium (brown millerite, larnite, calcium silicate oxide and calcite) and magnesium silicate, respectively [51]. Spectral bands in the range 1200–1000 cm^{-1} are due to stretching (in plane and out of plane) vibrations of Si-O groups [8,40,41].

3.7. *Testing of the Slag's Pozzolanic Activity*

The pozzolanic activity is one of the most important and basic parameters in order to determine if the investigated material can be used as a replacement for cement. For this reason, the AI of the slag was determined, and these investigations were performed in accordance with the EN 450-1 standard and the results are shown in Table 8.

Table 8. Test results of physical and mechanical characteristics of slag.

Properties	100% CEM	75% CEM + 25% Slag	Requirements EN 450-1
Activity index, %	-	28 days–76.9 90 days–76.4	28 days–75 * 90 days–85 *
Water Requirement, %	-	113	<95% of CEM *
Setting time, min	initial 115 final 165	initial 170 final 215	<2 × 100% CEM ** -
Soundness, mm	1.0	1.5	<10.0 **

* According to the EN 450-1 standard; ** according to EN 196-3.

Based on the results presented in Table 8, according to the parameters IST and S, the waste slag satisfied criteria, while according to parameters AI and WR, the waste slag does not satisfy parameters prescribed by the EN 450-1 standard, and thus, it may be concluded that the investigated slag does not possess pozzolanic activity in the presented form, and therefore it cannot be used as a replacement for cement without additional modifications and actions. In this study, this problem was solved with the addition of silica fume to the slag.

Namely, it is well known that silicate fume possesses good pozzolanic properties and because of this, it was tested as an addition to the slag in order to increase slag's pozzolanic activity and improve its properties. At first, determination of the chemical composition of the silica fume was performed and the results shown in Table 9 confirm its very good quality (≈94% of SiO_2).

Table 9. Chemical composition of silica fume expressed in weight percentages (wt.%).

Parameter	SiO_2	Fe_2O_3	Al_2O_3	CaO	MgO	SO_3	Na_2O	K_2O	P_2O_5	LOI
Silica fume	93.60	0.21	0.27	0.05	0.05	0.80	0.23	0.50	-	2.40

After the determination of the chemical composition, the silica fume was used as an addition to slag during the production of cement mixtures. For this aim, five samples of the cement mixtures were prepared with different amounts of CEM I 42.5 R (the content was in range 100–75%), slag (the content was in range 8.9–22.2%) and silica fume (the content was in range 1.1–2.8%) as described in Table 10.

After the mixtures were prepared, the AI, as the most important parameter of all prepared mixtures, were determined, and the results are presented in Table 11.

Table 10. Preparation of the mixtures with the additions of the different amounts of the silicate fume.

Mark	E	P10-3	P15-3	P20-3	P25-3
Description	Etalon (without CEM replacement)	CEM replacement with 8.9% of slag and 1.1% silica fume. Total replacement of CEM = 10%	CEM replacement with 13.3% of slag and 1.7% silica fume. Total replacement of CEM = 15%	CEM replacement with 17.8% of slag and 2.2% silica fume. Total replacement of CEM = 20%	CEM replacement with 22.2% of slag and 2.8% silica fume. Total replacement of CEM = 25%

Table 11. Activity indexes of mixtures with different amounts of silica fume.

Mix Code	Activity Index/%	Requirements EN 450-1/%
P10	28 days–95.1 90 days–93.2	28 days–75 90 days–85
P15	28 days–90.8 90 days–89.1	
P20	28 days–86.9 90 days–85.2	
P25	28 days–82.3 90 days–81.1	

From the presented results, it may be seen that slag with the addition of the silica fume possesses pozzolanic activity, which satisfies requirements prescribed by the EN 450-1 standard. Based on the measuring of the AI, the maximal possible cement replacement is 20% with 17.8% of slag and 2.2% of silica fume.

4. Conclusions

Results presented in this study show that waste slag—obtained from a heating plant (due to combustion of lignite) and taken from the landfill in Valjevo, Serbia—should be categorized as a hazardous waste due to its chemical composition and heavy metals content. From this point, it represents a serious environmental problem that must be solved.

Regarding slag utilization in the cement industry, results have shown that tested waste slag in the original form does not completely satisfy necessary criteria prescribed by the rulebooks from the aspect of the content of SiO_2, Al_2O_3 and Fe_2O_3 as well as from the aspect of the content of the unburned carbon and loss of ignition. The proposed solution and the easiest way for solving and overcoming these problems is thermal treatment of the slag prior to use. The results of the thermal analysis show that, after thermal treatment up to 650 °C, all criteria related to heavy metals content and chemical composition can be met. Additional physical–chemical characterizations (XRD, FTIR, N_2 physisorption, SEM, determination of the density) of the slag showed good correlation and that, regarding all additional tested physical–chemical properties, the slag completely satisfied the criteria prescribed by the rulebooks for utilization in cement industry.

Based on results of testing the physical–mechanical properties of the slag, it may be concluded that the investigated slag does not possess pozzolanic activity in its raw form, but with additional modifications and actions, it can be used as a replacement for cement. This may be overcome by mixing the slag with silica fume. The results show that mixing the slag with silica fume in minimal amounts (the slag/silica fume ration ≈0.88%/≈0.12%) improve the properties of the slag so that it possesses satisfactory pozzolanic activity. Maximal possible cement replacement was 20% with 17.8% of slag and 2.2% of silica fume. The results of leaching tests also prove that it is possible to use slag as mineral binders in cement composites, as there was no significant release of heavy metals.

Supplementary Materials: The following are available online at http://www.mdpi.com/2075-163X/10/11/0992/s1, Figure S1: Results of thermal analyses of the: (a) slag and (b) CEM I 42.5R. Figure S2: Nitrogen isotherm of slag sample (insert-part of BET isotherm for SBET calculation).

Author Contributions: M.K. conceived and designed the experiments, wrote the paper, and contributed to all experiments and the analyzing of the obtained results. A.N. contributed to all experiments and the analyzing of the obtained results. M.S. and J.G. participated in SEM and thermal analyses and analyzed the results. S.M. performed XRPD analyses and explained obtained results. J.K. performed the determination of textural properties and contributed to the analysis of obtained results. N.R. performed physical–mechanical measurements and contributed to analyzing the results. All authors have read and agreed to the published version of the manuscript.

Funding: This research received funding from the Ministry of Education and Science of Serbia.

Acknowledgments: This work has been supported by the Ministry of Education and Science of Serbia.

Conflicts of Interest: The authors declare no conflict of interest.

References

1. Mining Technology. Available online: https://www.mining-technology.com/features/feature-the-worlds-biggest-coal-reserves-by-country (accessed on 19 May 2020).
2. Milovanović, Z. Chapter: 7 Edition: Monographs "Energy and process plants". In *Thermal Power Plants—Theoretical Bases*; Faculty of Mechanical Engineering, University of Banja Luka: Banja Luka, Bosnia and Herzegovina, 2005; pp. 385–392.
3. American Geosciences Institute. Available online: https://www.americangeosciences.org/critical-issues/faq/what-are-the-different-types-of-coal (accessed on 19 May 2020).
4. Aydin, S.; Baradan, B. Effect of pumice and fly ash incorporation on high temperature resistance of cement based mortars. *Cem. Concr. Res.* **2007**, *37*, 988–995. [CrossRef]
5. Statista. Available online: https://www.statista.com/statistics/605024/major-lignite-consuming-countries-globally/date (accessed on 8 April 2020).
6. Dwivedi, A.; Jain, M. Fly ash—Waste management and overview: A Review. *Recent Res. Sci. Technol.* **2014**, *6*, 30–35.
7. Terzić, A.; Mijatović, N.; Milićević, L.J.; Radojević, Z. Slag from coal combustion process as a secondary raw material for use in building composites. In *Contemporary Materials and Constructions with Regulations, Belgrade, Serbia*; Jevtić, D., Ed.; Society for Testing and Research of Materials and Structures: Belgrade, Serbia, 2016; pp. 51–60.
8. Li, R.; Wang, L.; Yang, T.; Raninger, B. Investigation of MSWI fly ash melting characteristic by DSC–DTA. *Waste Manag.* **2007**, *27*, 1383–1392. [CrossRef] [PubMed]
9. Wei, M.C.; Wey, M.Y.; Hwang, J.H.; Chen, J.C. Stability of heavy metals in bottom ash and fly ash under various incinerating conditions. *J. Hazard. Mater.* **1998**, *57*, 145–154. [CrossRef]
10. Terzić, A.; Pezo, L.; Mijatović, N.; Stojanović, J.; Kragović, M.; Miličić, L.J.; Andrić, L.J. The effect of alternations in mineral additives (zeolite, bentonite, fly ash) on physico-chemical behavior of Portland cement based binders. *Constr. Build. Mater.* **2018**, *180*, 199–210. [CrossRef]
11. Sun, X.; Yi, Y. pH evolution during water washing of incineration bottom ash and its effect on removal of heavy metals. *Waste Manag.* **2020**, *104*, 213–219. [CrossRef] [PubMed]
12. European List of Waste, the Classification of Waste Based on: The European List of Waste (Commission Decision 2000/532/EC—Consolidated Version) and Annex III to Directive 2008/98/EC (Consolidated Version). Available online: http://data.europa.eu/eli/dec/2000/532/2015-06-01 (accessed on 25 September 2020).
13. Official Journal of the Republic of Serbia No. 12/95. Available online: http://www.slglasnik.com/ (accessed on 1 February 2017).
14. Decree of 20 December 2005—Official Journal of the Republic of Serbia No.113. Available online: http://www.slglasnik.com/ (accessed on 20 December 2005).
15. Liu, X.; Ni, C.; Meng, K.; Zhang, L.; Liu, D.; Sun, L. Strengthening mechanism of lightweight cellular concrete filled with fly ash. *Constr. Build. Mater.* **2020**, *251*, 118954. [CrossRef]
16. Wong, G.; Fan, X.; Gan, M.; Ji, Z.; Ye, H.; Zhou, Z.; Wang, Z. Resource utilization of municipal solid waste incineration fly ash in iron ore sintering process: A novel thermal treatment. *J. Clean. Prod.* **2020**, in press. [CrossRef]
17. Cai, J.; Pan, J.; Li, X.; Tan, J.; Li, J. Electrical resistivity of fly ash and metakaolin based geopolymers. *Constr. Build. Mater.* **2020**, *23420*, 117–868. [CrossRef]

18. Hu, C. Microstructure and mechanical properties of fly ash blended cement pastes. *Constr. Build. Mater.* **2014**, *73*, 618–625. [CrossRef]
19. Zabielska-Adamska, K. Laboratory compaction of fly ash and fly ash with cement additions. *J. Hazard. Mater.* **2008**, *151*, 481–489. [CrossRef]
20. Nadesan, M.; Dinakar, P. Mix design and properties of fly ash waste lightweight aggregates in structural lightweight concrete. *Case Stud. Constr. Mater.* **2017**, *7*, 336–347. [CrossRef]
21. Terzić, A.; Pavlović, L.J.; Radojević, Z.; Pavlović, V.; Mitić, V. Novel Utilization of Fly Ash for High-temperature Mortars: Phase Compositiiton, Microstructure and Performance Correlation. *Int. J. Appl. Ceram. Technol.* **2015**, *12*, 133–146. [CrossRef]
22. Wang, S.; Baxter, L.; Fonseca, F. Biomass fly ash in concrete: SEM, EDX and ESEM analysis. *Fuel* **2008**, *87*, 372–379. [CrossRef]
23. Adu-Amankwah, S.; Zajac, M.; Stabler, C.; Lothenbach, B. Influence of limestone on the hydration of ternary slag cements. *Cem. Concr. Res.* **2017**, *100*, 96–109. [CrossRef]
24. Sun, H.; Quian, J.; Yang, Y.; Fan, C.; Yue, Y. Optimization of gypsum and slag contents in blended cement containing slag. *Cem. Concr. Compos.* **2020**, *112*, 103674. [CrossRef]
25. Yoon, H.N.; Seo, J.; Kim, S.; Lee, H.K.; Park, S. Characterization of blast furnace slag-blended Portland cement for immobilization of Co. *Cem. Concr. Res.* **2020**, *134*, 106089. [CrossRef]
26. Voinovitch, I.; Debrad-Guedon, J.; Louvrier, J. *The Analysis of Silicates*; Israel Programfor Scientific Translations: Jerusalem, Israel, 1966.
27. Gregg, S.J.; Sing, K.S.W. *Adsorption, Surface Area and Porosity, Berichte der Bunsengesellschaft für Physikalische Chemie*; Hansen, N., Ed.; Academic Press: New York, NY, USA; London, UK, 1967; p. 71.
28. Dubinin, M.M. Physical adsorption. In *Progress in Surface Press, and Membrane Science*; Cadenead, D.A., Danielli, J.F., Rosenberg, M.D., Eds.; Academic Press: New York, NY, USA; London, UK, 1975; Volume 9, pp. 1–70.
29. Dollimore, D.; Heal, G.R. An improved distribution method for the calculation of pore size from adsorption data. *J. Appl. Chem.* **1964**, *14*, 109–114. [CrossRef]
30. Hewlett, P.C. *Lea's Chemistry of Cement and Concrete*, 4th ed.; Elsevier Ltd.: Oxford, UK, 1998.
31. Marinković, S.; Trifunović, P.; Tokalić, R.; Matijašević, S.; Kostić-Pulek, A. DTA/TGA studies of bottom ash from the Nikola Tesla power plant from Serbia for the purpose of its utilization in road construction. In Proceedings of the 36th International Conference of SSCHE, Tatranské Matliare, Slovakia, 25–29 May 2009; pp. 094-1–094-8.
32. Maroto-Valer, M.; Taulbee, D.; Hower, J. Characterization of differing forms of unburned carbon present in fly ash separated by density gradient centrifugation. *Fuel* **2001**, *80*, 759–800. [CrossRef]
33. Hill, R.; Sarkar, S.; Rathbone, R.; Hower, J. An examination of fly ash carbon and itsinteractions with air entraining agent. *Cem. Concr. Res.* **1997**, *27*, 193–204. [CrossRef]
34. Millán-Corrales, G.; González-López, J.R.; Palomo, A.; Fernandez-Jiménez, A. Replacing fly ash with limestone dust in hybrid cements. *Constr. Build. Mater.* **2020**, *243*, 118–169. [CrossRef]
35. Soriano, I.; Tashima, M.; Bonilla, M.; Payá, J.; Monzó, J.; Borrachero, M. Use of high resolution thermo gravimetric analysis (HRTG) technique in spent FCC catalyst/Portland cement pastes. *J. Therm. Anal. Calorim.* **2015**, *120*, 1511–1517. [CrossRef]
36. Scrivener, K.; Snellings, R.; Lothenbach, B. *Chapter 5 (Thermogravimetric Analysis), A Practical Guide to Micostructural Analysis of Cementitious Materials*, 1st ed.; Taylor and Francis Group: Oxford, UK, 2016; p. 36.
37. Zeng, Q.; Li, K.; Chong, T.; Dangla, P. Determination of cement hydration and pozzolanic reaction extents for fly-ash cement pastes. *Constr. Build. Mater.* **2012**, *27*, 560–569. [CrossRef]
38. Bertier, P.; Schweinar, K.; Stanjek, H.; Ghanizadeh, A.; Clarkson, C.R.; Busch, A.; Kampman, N.; Prinz, D.; Amann-Hildenbrabd, A.; Krooss, B.M.; et al. On the use and abuse of N_2 physisorption for the characterization of the pore structure of shales. *Clay Miner. Soc. Workshop Lect.* **2016**, *21*, 151–161.
39. Thommes, M.; Kaneko, K.; Neimark, V.A.; James, P.O.; Rodriguez-Reinoso, F.; Rouquerol, J.; Sing, W.K.S. Physisorption of gases, with special reference to the evaluation of surface area and pore size distribution (IUPAC Technical Report). *Pure Appl. Chem.* **2015**, *87*, 1051–1069. [CrossRef]
40. Patel, H.A.; Somani, S.R.; Bajaj, C.H.; Jasra, V.R. Nanoclays for polymer nanocomposites, paints, inks, greases and cosmetics formulations, drug delivery vehicle and waste water treatment. *Bull. Mater. Sci.* **2006**, *29*, 133–145. [CrossRef]

41. Wu, H.; Xie, H.; He, G.; Guan, Y.; Zhang, Y. Effects of the pH and anions on the adsorption of tetracycline on iron-montmorillonite. *Appl. Clay Sci.* **2016**, *119*, 161–169. [CrossRef]
42. Fesenko, O.; Yatsenko, L.; Brodin, M. *Nano materials Imaging Techniques, Surface Studies, and Applications*; Springer: Berlin, Germany, 2012; p. 146.
43. Database of ATR-FT-IR Spectra of Various Materials. Available online: http://lisa.chem.ut.ee/IR_spectra/paint/fillers/quartz/ (accessed on 23 May 2020).
44. Correcher, V.; Garcia-Guinea, J.; Bustillo, A.M.; Garcia, R. Study of the thermoluminescence emission of a natural α-cristobalite. *Radiat. Eff. Defects Solids* **2009**, *164*, 59–67. [CrossRef]
45. Bosch-Reig, F.; Gimeno-Adelantado, J.V.; Bosch-Mossi, F.; Doménech-Carbó, A. Quantification of minerals from ATR-FTIR spectra with spectral interferences using the MRC method. *Spectrochim. Acta Part A Mol. Biomol. Spectrosc.* **2017**, *181*, 7–12. [CrossRef]
46. Obradović, M. Research and Comparison of the Influence of Low Rank Coals Characteristics on Their Grindability and Milling Process Parameters. Ph.D. Thesis, Faculty of Mechanical Engineering, University of Belgrade, Belgrade, Serbia, 2015.
47. Mitrović, D. Origin and Paleoenvironmental Reconstruction of the "Kovin" Coal Deposit—Implications from Petrographic and Geochemical Investigations. Ph.D. Thesis, Faculty of Chemistry, University of Belgrade, Belgrade, Serbia, 2018.
48. Mayo, D.V.; Miller, F.A.; Hannah, R.W. *Course Notes on the Interpretation of Infrared and Raman Spectra*; John Wiley and Sons: Hoboken, NJ, USA, 2003.
49. Mane, V.G. Studies of Some Transition Metal Complexes of Schiffõs Bases. Lulu.Com. 2019. Available online: https://books.google.rs/books?hl=en&lr=&id=RVePDwAAQBAJ&oi=fnd&pg=PA19&ots=_Z8TVoH9Ph&sig=U2YqsX98CeXlMr4exwqcajZTtok&redir_esc=y#v=onepage&q&f=false (accessed on 25 September 2020).
50. Singh, S.; Menon, S.; Gupta, K.; Jayavel, R. Preferentially oriented single crystal growth of brownmillerite $CaFeO_{2.5}$ by flux growth technique. *Mater. Lett.* **2014**, *131*, 332–335. [CrossRef]
51. Choudhary, R.; Koppala, S.; Swamiappan, S. Bioactivity studies of calcium magnesium silicate prepared from eggshell wastebysol–gelcombustionsynthesis. *J. Asian Ceram. Soc.* **2015**, *3*, 173–177. [CrossRef]

minerals

MDPI

Article

Prediction Models for Evaluating the Strength of Cemented Paste Backfill: A Comparative Study

Jiandong Liu, Guichen Li, Sen Yang and Jiandong Huang *

School of Mines, China University of Mining and Technology, Xuzhou 221116, China;
tb16020009b0@cumt.edu.cn (J.L.); liguichen@cumt.edu.cn (G.L.); TS20020064A31TM@cumt.edu.cn (S.Y.)
* Correspondence: huang@cumt.edu.cn

Received: 24 October 2020; Accepted: 17 November 2020; Published: 21 November 2020

Abstract: Cemented paste backfill (CPB) is widely used in underground mining, and attracts more attention these years as it can reduce mining waste and avoid environmental pollution. Normally, to evaluate the functionality of CPB, the compressive strength (UCS) is necessary work, which is also time and money consuming. To address this issue, seven machine learning models were applied and evaluated in this study, in order to predict the UCS of CPB. In the laboratory, a series of tests were performed, and the dataset was constructed considering five key influencing variables, such as the tailings to cement ratio, curing time, solids to cement ratio, fine sand percentage and cement types. The results show that different variables have various effects on the strength of CPB. The optimum models for predicting the UCS of CPB are a support vector machine (SVM), decision tree (DT), random forest (RF) and back-propagation neural network (BPNN), which means that these models can be directly applied for UCS prediction in future work. Furthermore, the intelligent model reveals that the tailings to cement ratio has the most important influence on the strength of CPB. This research can boost CPB application in the field, and guide the artificial intelligence application in future mining.

Keywords: cemented paste backfill; beetle antennae search; support vector machine; sensitive analysis; evaluation

1. Introduction

Cemented paste backfill (CPB) is an environment-friendly material, as it uses mining waste as the main aggregate [1–5]. Normally, cement and water should be added to CPB to increase the strength and flowability. The function of CPB is to fill the underground stope and to keep the work environment safe. Due to its key role of support of underground space in mines, it attracts more attention these years, especially in research areas such as strength and durability, etc. [6–9].

As described above, the compressive strength is a key indicator to evaluate the functionality of CPB in the underground mine, and hence most studies focus on the experiments and revealing the mechanism. Normally, the strength of CPB was tested in the laboratory, and mine engineers find the best proportion of materials in CPB [8,10,11]. However, experimental studies are time and money consuming, and are not suitable for a mass of experiments [12]. Parts of scholars apply empirical formulas to predict the strength of CPB, and parts of them are useful, but most of the formulas are constrained by its inputs and dataset [13,14]. Due to the low accuracy of empirical formulas, there is a limited application of the formulation [15,16]. To predict the strength of CPB accurately, more simple and reliable models should be presented and utilized.

It is a hot research interest in the concrete area of using artificial intelligence approaches to evaluate or predict mechanical properties [17–20]. Furthermore, some researchers have applied machine learning methods to assess the properties of CPB. For example, the artificial neural network (ANN) and other evolved optimization algorithms have been applied to predict the compressive strength of

CPB [21]. Moreover, more intelligent algorithms such as random forest (RF), support vector machine (SVM) and decision trees (DT) have been utilized in such areas [22]. It should be pointed out that most machine learning models can make the prediction accurately, but they are limited studies on the effect of different models on the accuracy of prediction. The mentioned algorithms are differently sensitive for various datasets with different properties. Therefore, it is necessary to use different models to predict the compression and find the optimum model for future applications [23].

In this study, various experimental tests were conducted to construct the dataset considering five key factors: the cement type; cement-coarse tailings ratio; solids-water ratio; fine tailings percentage; and time. Then, eight machine learning models, i.e., BPNN, SVM, DT and RF, were used for the prediction of the strength of CPB, and the hyperparameters of these models were tuned by a beetle antennae search algorithm (BAS). Lastly, the prediction results were compared and analyzed, which give the best models in CPB strength prediction. Furthermore, the importance of influencing variables was conducted, and the most sensitive factor was found. The study provides the best models for CPB strength prediction, which is an important guideline in this research field.

2. Materials and Experimental Methods

2.1. Experimental Tests

In this field, the CPB was normally composited with water, different cement types, coarse tailings and fine tailings. In this study, the CPB specimen was constructed in the laboratory. The component of coarse tailings and fine tailings are determined by X-ray diffraction (XRD), and the results are given in Figure 1. The mineralogical composition of coarse tailings and fine tailings are given in Tables 1 and 2, respectively. The particle size distribution of fine tailings and coarse tailings is given in Figure 2. As for the Portland cement PO. 325R, two types of cement (1# (Shangfeng cement), 2# (Maan Cement)) were selected as a binder, and a series of tests were performed. The test results are shown in Table 3. The water chemical elements were analyzed by ICP, and the results are given in Table 4. Importantly, the scheme of a mix of CPB is summarized in Table 5. In China, to obtain a relatively high strength of CPB to keep the unground mine safe, more cement is normally added. According to the previous study [21], the coarse tailings-cement ratio was set as 4, 6, 8 and 10. Due to the negative effect of fine tailings percentage on CPB strength, the percentage of fine tailings was selected as 0%, 10%, 15% and 20%. Considering the fluidity of CPB slurry, the solids-water ratio should be under 72%, and then the ratio was set as 68%, 70% and 72% in this study. Curing time is also an important factor affecting the strength of CPB, and it is therefore chosen as seven days, 28 days and 60 days.

A mixer (UJZ-15) was used for mixing cement, tailings and water. First, the tailings and cement were mixed for 0.5 min, and then with the increase of time, the water was poured into the mixture. The water, cement and tailings were continuously mixed for at least five minutes until the components mixed well. Then, the mixture was poured into the cubical molds (70.1 mm × 70.1 mm × 70.1 mm) and tamped uniformly. After 48 h, the CPB specimen was demolded. Furthermore, the test specimen was placed at a constant temperature (20 °C) and humidity (90%) for curing (Figure 3a).

A total of 870 specimens were constructed, and were used for obtaining the compressive strength by a hydraulic servo testing machine (Figure 3b), according to Chinese standards (GBT50107-2019 [24]). The standard deviations of UCS test results are given in Table 6. It should be pointed out that the UCS test should normally use ASTM standard, and therefore the cylindrical specimens should be prepared. However, the cubical sample is easier for casting and demolding. Furthermore, a large number of specimens need to find a better way to improve efficiency. Hence, the cubical specimens were applied in this study. According to research by R.L. Hermite, there is a conversion formula between the strength of the cylindrical specimen and the strength of the cubical specimen, as follows:

$$f_{cylider} = 1.05 f_{cube}[0.76 + 0.2 \times \log(\frac{6.89 \times 10^{-3} f_{cube}}{2840})] \quad (1)$$

where $f_{cylinder}$ is the strength of cylindrical specimen (diameter of 150 mm and height of 300 mm, Unit: MPa); f_{cube} is the strength of cubical specimen (side length 70.1 mm, Unit: MPa).

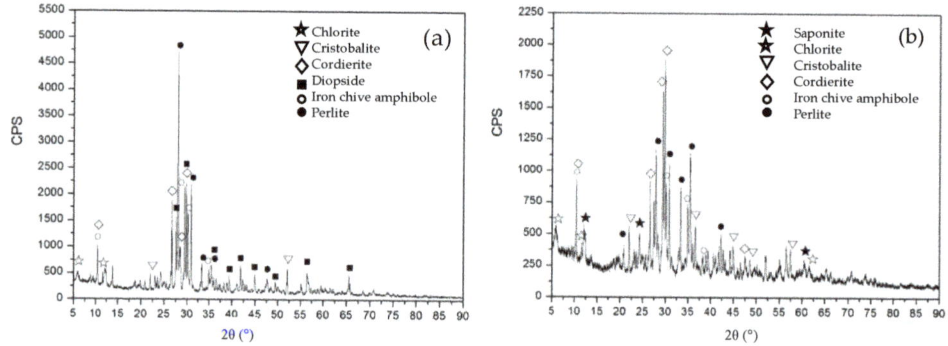

Figure 1. Results of X-ray diffraction (XRD) for coarse tailings and fine tailings: (**a**) coarse tailings; (**b**) fine tailings.

Table 1. Mineralogical composition of coarse tailings.

Elements	SiO_2	Al_2O_3	TFe	FeO	MgO	CaO	Na_2O
Content %	46.52	6.2	7.86	5.44	8.99	19.2	1.64
Elements	K_2O	TiO_2	P_2O_5	MnO	S	Cu	Zn
Content %	0.25	0.27	0.18	0.21	0.61	0.11	5.08

Table 2. Mineralogical composition of fine tailings.

Elements	SiO_2	Al_2O_3	TFe	FeO	MgO	CaO	Na_2O
Content %	43.4	5.67	8.56	7.43	10.65	18.89	1.44
Elements	K_2O	TiO_2	P_2O_5	MnO	S	Cu	Zn
Content %	0.31	0.25	0.25	0.19	0.94	0.2	6.14

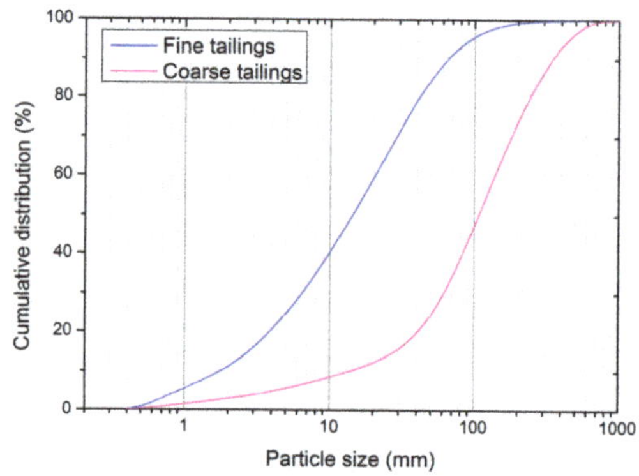

Figure 2. The particle size distribution of fine tailings and coarse tailings.

Table 3. The tests results of different cement.

Sample	Fineness (<0.0045 mm/%)	Initial Setting Time/min	Final Setting Time/min	28d-UCS/MPa	28d-Flexural Strength/MPa
#1	23.2	240	305	30.7	6.5
#2	6	180	255	39.8	8

Table 4. Results of chemical element analysis of water.

Program	Results/(mg/L)	Program	Results/(mg/L)
Al	0.39	Li	<0.05
As	<0.05	Mg	20.8
Ba	0.058	Mn	0.22
Be	<0.05	Ni	<0.05
Bi	0.074	Pb	<0.05
Ca	6.26	Sb	<0.05
Cd	<0.05	Sn	<0.05
Co	<0.05	Sr	6.74
Cr	<0.05	Ti	<0.05
Cu	<0.05	V	<0.05
Fe	0.15	Zn	<0.05

Table 5. The experimental design of CPB.

Cement Types	Coarse Tailings-Cement Ratio	Fine Tailings Percentage	Solids-Water Ratio	Curing Time (Days)
#1	4	0%	0.68	7
#2	6	10%	0.70	28
	8	15%	0.72	60
	10	20%		

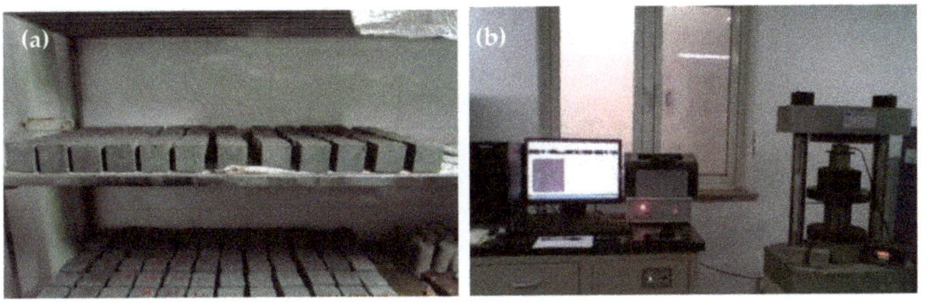

Figure 3. CPB specimen and its UCS test: (**a**) Prepared specimen; (**b**) UCS tests.

Table 6. The standard deviations of UCS test results.

Cement Type	Curing Time (Days)	Standard Deviations of UCS
	7	0.47
#1	28	0.81
	60	0.85
	7	0.57
#2	28	0.96
	60	0.94

2.2. Applied Machine Learning Models

In this study, seven machine learning models, i.e., back-propagation neural network (BPNN), support vector machine (SVM), decision tree (DT), random forest (RF), k-nearest neighbors (KNN), mixed logistic regression (MLR) and logistic regression (LR) were utilized for the prediction of the UCS of CPB [25–29]. In these models, the input variables were set as cement types, curing time, coarse tailings-cement ratio, fine tailings percentage and solids-water ratio, while the UCS is the output. These machine learning algorithms are widely used in regression problems, such as the prediction of concrete properties, prediction of rock properties and prediction of rockburst events. However, these models have not been compared, especially in CPB materials. Thus, it is necessary to compare them and find the best model to predict the UCS of CPB.

2.3. Building Waste Heat Simulation with EnergyPlus

As for the performance of a machine learning model, the evaluation method is critical, as it can help the researcher to find the optimum algorithm for application. In this study, the dataset was split into two parts i.e., training set (70%) and testing set (30%) [30]. Then, the 10-fold cross-validation approach was used for validation in each dataset. According to the literature, the correlation coefficient (R) and root-mean-square error (RMSE) are utilized in this study for assessing the performance of the applied model and the formula are as follows:

$$R = \frac{\sum_{i=1}^{N}(y_i^* - \overline{y^*})(y_i - \overline{y})}{\sqrt{\sum_{i=1}^{N}(y_i^* - \overline{y^*})^2}\sqrt{\sum_{i=1}^{N}(y^i - \overline{y})^2}} \tag{2}$$

$$\text{RMSE} = \sqrt{\frac{1}{N}\sum_{i=1}^{N}(y_i^* - y_i)^2} \tag{3}$$

where N is the numbers in the dataset; y_i^* and y_i are the expected values and real values, respectively; $\overline{y}$ and $\overline{y^*}$ means the mean predicted values and mean actual values, respectively.

3. Results and Discussion

3.1. The Results of UCS of Cemented Paste Backfill (CPB)

Based on the experimental results, all data are given in Figure 4. It is evident from Figure 4 that, normally, fine tailing percentage plays a negative effect on the cemented paste backfill (CPB) strength, which means that with the decrease of fine tailing, the UCS of CPB increases. However, it should be pointed out that in some cases, the effect of fine tailing is not clear. Moreover, with the increase of curing time, the strength of CPB increases quickly. A similar trend can be found in the solids to water ratio, and the UCS increases with the increase of that percentage. However, the strength decreases with the increase of the tailing to cement ratio. Cement types also affect the strength of CPB. The CPB with #2 cement has a higher strength than that with cement #1. The results obtained from the laboratory are consistent with the literature mentioned before, indicating that the dataset is reliable and accurate.

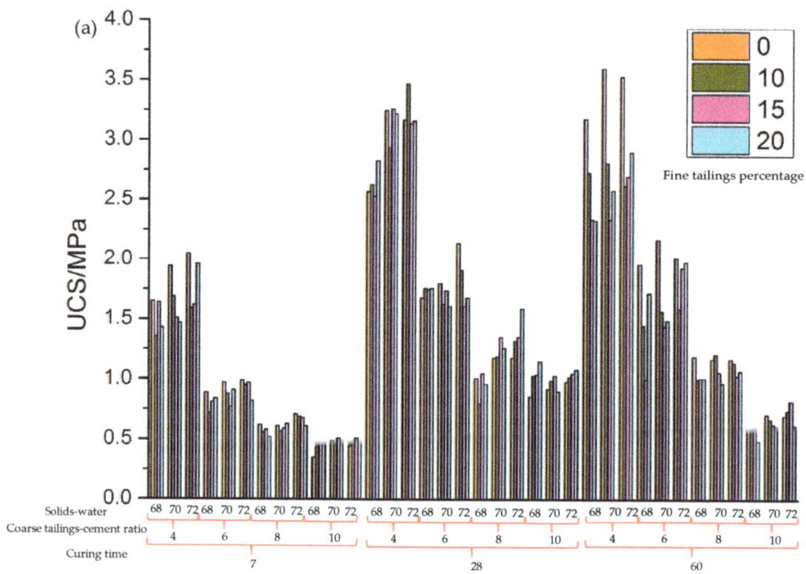

Figure 4. *Cont.*

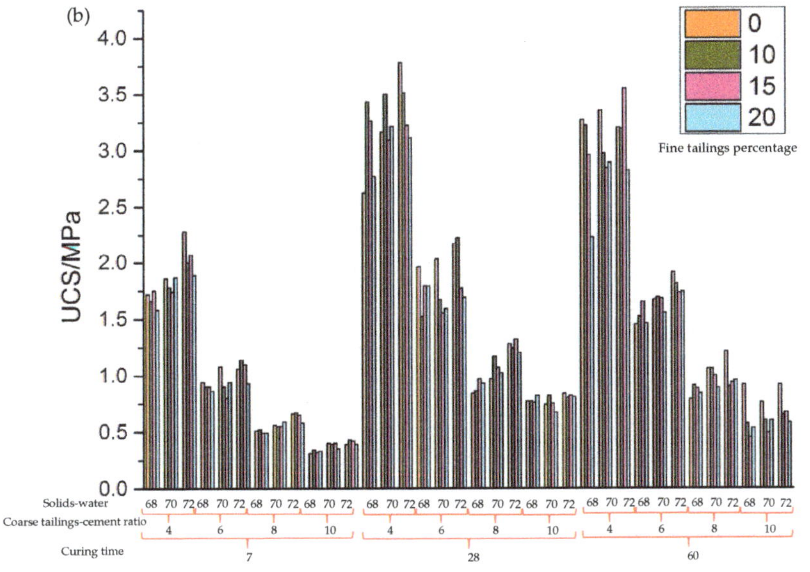

Figure 4. UCS test results of CPB in 4D. (**a**) 1# cement; (**b**) 2# cement.

3.2. *Hyper-Parameters Tuning*

For different machine learning models, the BAS is used for finding the optimum hyperparameters. As can be seen from Figure 5, the RMSE decreased quickly after several iterations, and then the value was kept stable, which means that the BAS can tune these models effectively. Moreover, from Figure 5, it is clear that the SVM model can achieve the smallest RMSE after 10 iterations due to the BAS tuning, while the BAS is inefficient in the LR and MLR models.

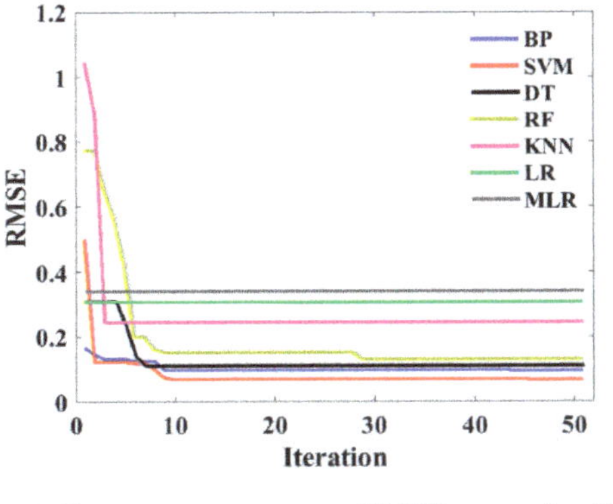

Figure 5. Root-mean-square error (RMSE) versus iterations.

3.3. *Model Evaluation*

The seven models were systematically compared, as shown in Figure 6. The root-mean-square error (RMSE) and R for the training dataset and testing dataset were given in Figure 6. It can be seen that the support vector machine (SVM), random forest (RF), back-propagation (BP) and decision tree (DT) models show better performance than other models. It should be pointed out that for different datasets, the performance of the same model shows different predictive accuracy. For example, k-nearest neighbors (KNN) has the best performance on the training dataset, with the highest R (0.9963) and the lowest RMSE (0.0774), although the KNN model shows a not perfect performance on the testing set, with a low R (0.8992) and a high RMSE (0.3682). As for the LR and MLR, it can be observed that the prediction performance is worse compared with other machine learning models, which indicates that the baseline models are inefficient in this CPB dataset.

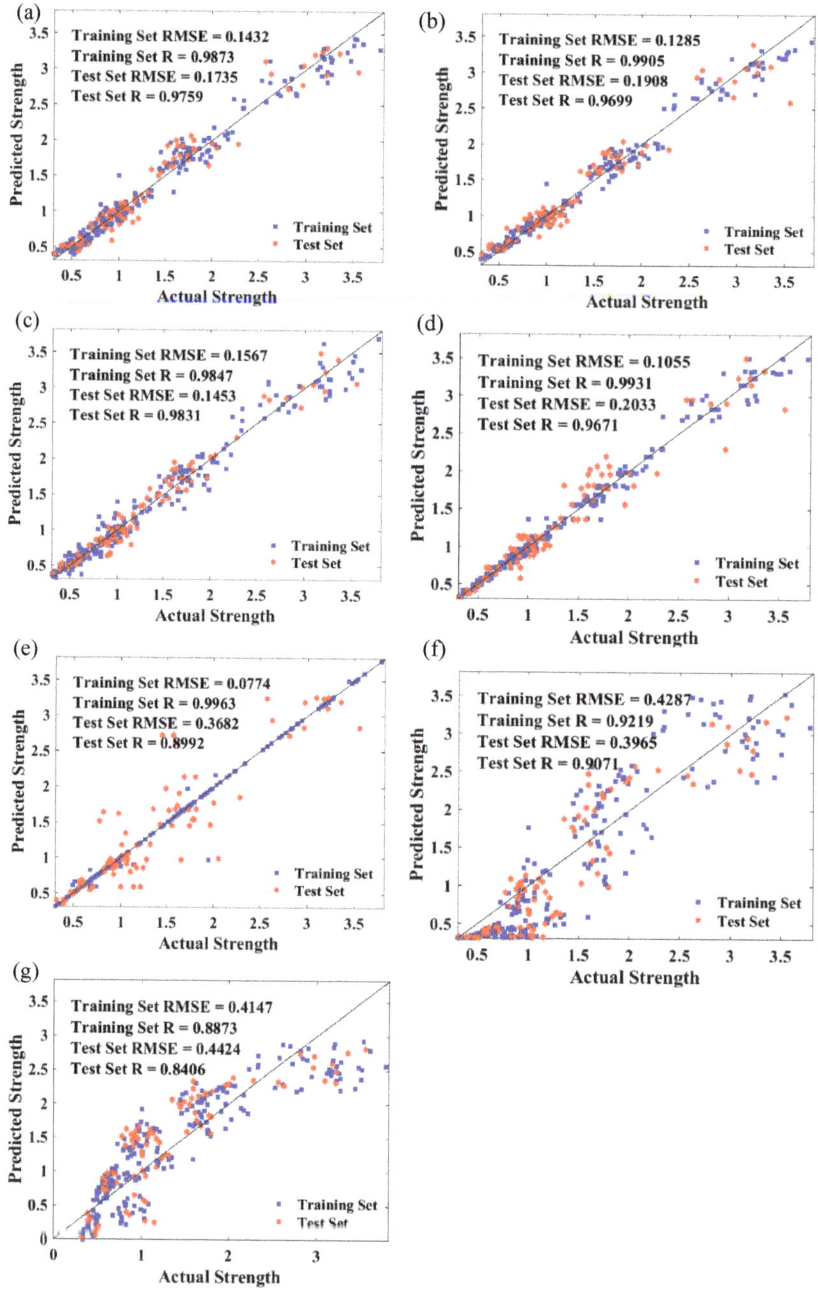

Figure 6. Comparison of different models: (**a**) support vector machine (SVM); (**b**) random forest (RF); (**c**) back-propagation (BP); (**d**) decision tree (DT); (**e**) k-nearest neighbors (KNN); (**f**) logistic regression (LR); and (**g**) mixed logistic regression (MLR).

3.4. Spatial Variations of Anthropogenic Heat Intensity

The UCS of CPB can be effectively predicted by the proposed model, and then the relative importance of variables can also be accurately determined by machine learning models. In the study, the RF model was used for calculating variable importance, and the results were presented in Figure 7. It is evident that the ratio between tailings and cement is the most important variable over the others, which has an influencing score of 7.06. Curing time (5.593) also works as an important parameter in the UCS of CPB, followed by the solids to water ratio (0.85) and fine sand. However, the cement type (0.413) has the least importance on the strength of CPB, which means that engineers should pay more attention to cement content. Furthermore, adding more cement can obviously increase the CPB strength. The above results are quite useful for CPB application, and can guide the field study.

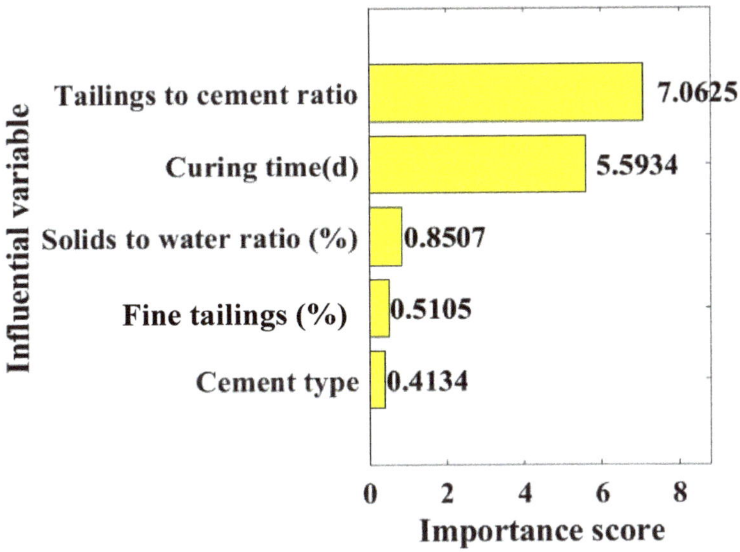

Figure 7. Variable importance of CPB.

4. Conclusions

Based on a series of experimental studies of CPB, the dataset was constructed, and was used for evaluating the proposed models. Seven machine learning models were applied for predicting the strength of CPB, and then different models were compared. The relative results are as follows:

1. The results of UCS of CPB shows that with the increase of ratio between solids and water ratio, as well as curing time, the UCS of CPB increased, while the strength of CPB declined with the increase of fine sand percentage and the tailing to cement ratio.
2. According to the prediction models, the SVM RF, BP and DT models can predict the UCS of CPB effectively and accurately, although the KNN, LR and MLR have a relatively worse performance on the prediction.
3. The tailing to cement ratio can affects the strength of CPB obvious, followed by Curing time, solids to water ratio, fine sand, and cement type, which can guide the CPB application in the field.

It should be pointed out that the accuracy and reliability of machine learning algorithms depend on the dataset, i.e., the number and types of samples. Thus, in the future, more samples should be obtained and more efficient models should be proposed.

Author Contributions: Conceptualization, J.L.; methodology, J.L. and J.H.; writing—original draft preparation, J.L. and S.Y.; writing—review and editing, G.L.; supervision, J.H. All authors have read and agreed to the published version of the manuscript.

Funding: This study was financially supported by the "Fundamental Research Funds for the Central Universities (grant number: 2020ZDPY0221)".

Conflicts of Interest: The authors declare no conflict of interest.

References

1. Qi, C.; Fourie, A.; Chen, Q.; Zhang, Q. A strength prediction model using artificial intelligence for recycling waste tailings as cemented paste backfill. *J. Clean. Prod.* **2018**, *183*, 566–578. [CrossRef]
2. Chen, Q.-S.; Zhang, Q.-L.; Fourie, A.; Chen, X.; Qi, C.-C. Experimental investigation on the strength characteristics of cement paste backfill in a similar stope model and its mechanism. *Constr. Build. Mater.* **2017**, *154*, 34–43. [CrossRef]
3. Qi, C.; Fourie, A. Cemented paste backfill for mineral tailings management: Review and future perspectives. *Miner. Eng.* **2019**, *144*, 106025. [CrossRef]
4. Belem, T.; Benzaazoua, M.; Bussière, B. Mechanical behaviour of cemented paste backfill. In Proceedings of the 53rd Canadian Geotechnical Conference, Montreal, QC, Canada, 15–18 October 2000; pp. 373–380.
5. Benzaazoua, M.; Bussière, B.; Demers, I.; Aubertin, M.; Fried, É.; Blier, A. Integrated mine tailings management by combining environmental desulphurization and cemented paste backfill: Application to mine doyon, Quebec, Canada. *Miner. Eng.* **2008**, *21*, 330–340. [CrossRef]
6. Qi, C.; Tang, X.; Dong, X.; Chen, Q.; Fourie, A.; Liu, E. Towards intelligent mining for backfill: A genetic programming-based method for strength forecasting of cemented paste backfill. *Miner. Eng.* **2019**, *133*, 69–79. [CrossRef]
7. Qi, C.; Chen, Q.; Fourie, A.; Zhang, Q. An intelligent modelling framework for mechanical properties of cemented paste backfill. *Miner. Eng.* **2018**, *123*, 16–27. [CrossRef]
8. Fall, M.; Pokharel, M. Coupled effects of sulphate and temperature on the strength development of cemented tailings backfills: Portland cement-paste backfill. *Cem. Concr. Compos.* **2010**, *32*, 819–828. [CrossRef]
9. Komurlu, E.; Kesimal, A.; Demir, S. Experimental and numerical analyses on determination of indirect (splitting) tensile strength of cemented paste backfill materials under different loading apparatus. *Geomech. Eng.* **2016**, *10*, 775–791. [CrossRef]
10. Sun, Y.; Li, G.; Zhang, J.; Sun, J.; Xu, J. Development of an ensemble intelligent model for assessing the strength of cemented paste backfill. *Adv. Civ. Eng.* **2020**, *2020*, 5198583. [CrossRef]
11. Helinski, M.; Fahey, M.; Fourie, A. Behavior of cemented paste backfill in two mine stopes: Measurements and modeling. *J. Geotech. Geoenviron.* **2011**, *137*, 171–182. [CrossRef]
12. Mashrei, M.A.; Seracino, R.; Rahman, M. Application of artificial neural networks to predict the bond strength of frp-to-concrete joints. *Constr. Build. Mater.* **2013**, *40*, 812–821. [CrossRef]
13. González-Taboada, I.; González-Fonteboa, B.; Martínez-Abella, F.; Pérez-Ordóñez, J.L. Prediction of the mechanical properties of structural recycled concrete using multivariable regression and genetic programming. *Constr. Build. Mater.* **2016**, *106*, 480–499. [CrossRef]
14. Thompson, B.D.; Bawden, W.F.; Grabinsky, M. In situ measurements of cemented paste backfill at the cayeli mine. *Can. Geotech. J.* **2012**, *49*, 755–772. [CrossRef]
15. Sun, Y.; Li, G.; Zhang, J. Developing hybrid machine learning models for estimating the unconfined compressive strength of jet grouting composite: A comparative study. *Appl. Sci.* **2020**, *10*, 1612. [CrossRef]
16. Kolovos, K.G.; Asteris, P.G.; Tsivilis, S. Properties of sandcrete mixtures modified with metakaolin. *Eur. J. Environ. Civ. Eng.* **2016**, *20*, s18–s37. [CrossRef]
17. Ouellet, S.; Bussière, B.; Aubertin, M.; Benzaazoua, M. Microstructural evolution of cemented paste backfill: Mercury intrusion porosimetry test results. *Cem. Concr. Res.* **2007**, *37*, 1654–1665. [CrossRef]
18. Zhang, J.; Li, D.; Wang, Y. Toward intelligent construction: Prediction of mechanical properties of manufactured-sand concrete using tree-based models. *J. Clean. Prod.* **2020**, *258*, 120665. [CrossRef]
19. Sun, J.; Zhang, J.; Gu, Y.; Huang, Y.; Sun, Y.; Ma, G. Prediction of permeability and unconfined compressive strength of pervious concrete using evolved support vector regression. *Constr. Build. Mater.* **2019**, *207*, 440–449. [CrossRef]

20. Sun, Y.; Li, G.; Zhang, J.; Qian, D. Prediction of the strength of rubberized concrete by an evolved random forest model. *Adv. Civ. Eng.* **2019**, *2019*, 1643529. [CrossRef]
21. Zhang, J.; Huang, Y.; Ma, G.; Sun, J.; Nener, B. A metaheuristic-optimized multi-output model for predicting multiple properties of pervious concrete. *Constr. Build. Mater.* **2020**, *249*, 118803. [CrossRef]
22. Le, L.T.; Nguyen, H.; Dou, J.; Zhou, J. A comparative study of pso-ann, ga-ann, ica-ann, and abc-ann in estimating the heating load of buildings' energy efficiency for smart city planning. *Appl. Sci.* **2019**, *9*, 2630. [CrossRef]
23. Sun, Y.; Zhang, J.; Li, G.; Wang, Y.; Sun, J.; Jiang, C. Optimized neural network using beetle antennae search for predicting the unconfined compressive strength of jet grouting coalcretes. *Int. J. Numer. Anal. Methods Geomech.* **2019**, *43*, 801–813. [CrossRef]
24. GBT50107-2019. *Standard for Inspection and Evaluation of Concrete Strength*; Ministry of urban and rural construction and environmental protection of the people's Republic of China: Beijing, China, 2019.
25. Sun, Y.; Zhang, J.; Li, G.; Ma, G.; Huang, Y.; Sun, J.; Wang, Y.; Nener, B. Determination of young's modulus of jet grouted coalcretes using an intelligent model. *Eng. Geol.* **2019**, *252*, 43–53. [CrossRef]
26. Qi, C.; Fourie, A.; Chen, Q. Neural network and particle swarm optimization for predicting the unconfined compressive strength of cemented paste backfill. *Constr. Build. Mater.* **2018**, *159*, 473–478. [CrossRef]
27. Qi, C.-C. Big data management in the mining industry. *Int. J. Miner. Metall. Mater.* **2020**, *27*, 131–139. [CrossRef]
28. Sun, Y.; Li, G.; Zhang, N.; Chang, Q.; Xu, J.; Zhang, J. Development of ensemble learning models to evaluate the strength of coal-grout materials. *Int. J. Min. Sci. Technol.* **2020**. [CrossRef]
29. Zhang, J.; Huang, Y.; Aslani, F.; Ma, G.; Nener, B. A hybrid intelligent system for designing optimal proportions of recycled aggregate concrete. *J. Clean. Prod.* **2020**, *273*, 122922. [CrossRef]
30. Zhang, J.; Huang, Y.; Wang, Y.; Ma, G. Multi-objective optimization of concrete mixture proportions using machine learning and metaheuristic algorithms. *Constr. Build. Mater.* **2020**, *253*, 119208. [CrossRef]

Publisher's Note: MDPI stays neutral with regard to jurisdictional claims in published maps and institutional affiliations.

minerals

MDPI

Article

Stepwise Utilization Process to Recover Valuable Components from Copper Slag

Siwei Li, Zhengqi Guo *, Jian Pan, Deqing Zhu, Tao Dong and Shenghu Lu

School of Minerals Processing and Bioengineering, Central South University, Changsha 410083, China; swli@csu.edu.cn (S.L.); pjcsu@csu.edu.cn (J.P.); dqzhu@csu.edu.cn (D.Z.); dongtao@csu.edu.cn (T.D.); lushenghu@csu.edu.cn (S.L.)

* Correspondence: guozqcsu@csu.edu.cn; Tel.: +86-731-8883-6041

Abstract: Waste copper slag is a typical hazardous solid waste containing a variety of valuable elements and has not been effectively disposed of so far. In this paper, a stepwise extraction process was proposed to recover valuable elements (copper, iron, lead and zinc) from waste copper slag. The specific procedures are as follows: (1) A flotation process was adopted to enrich copper, and when the copper grade in the flotation concentrate was 21.50%, the copper recovery rate was 77.78%. (2) The flotation tailings were pelletized with limestone, then the green pellets were reduced, and the magnetic separation process was carried out. When the iron and copper grades in the magnetic concentrate were 90.21% Fe and 0.4% Cu, 91.34% iron and 83.41% copper were recovered, respectively. (3) Non-magnetic tailings were mixed with clinker and standard sand to produce common Portland cement. Several products were obtained from the waste copper slag through the proposed process: flotation concentrate, measured 21.50% Cu; magnetic concentrate, containing 90.21% TFe and 0.4% Cu; direct reduction dust, including 65.17% ZnO and 2.66% PbO; common Portland cement for building construction. The comprehensive utilization method for waste copper slag achieved zero tailing and has great potential for practical application.

Keywords: waste copper slag; flotation; direct reduction-magnetic separation; common Portland cement

Citation: Li, S.; Guo, Z.; Pan, J.; Zhu, D.; Dong, T.; Lu, S. Stepwise Utilization Process to Recover Valuable Components from Copper Slag. *Minerals* **2021**, *11*, 211. https://doi.org/10.3390/min11020211

Academic Editor: Kevin Galvin

Received: 18 December 2020
Accepted: 8 February 2021
Published: 17 February 2021

1. Introduction

One of the greatest challenges in the mining and processing industries is waste management and waste copper slag is defined as one such waste [1]. Waste copper slag (WCS) is a by-product produced in the copper pyrometallurgical process. It is estimated that 30 million tons of WCS are generated and dumped globally every year [2,3]. In addition, the slag contains a lot of harmful elements, such as Cu, Pb and Zn, which cause pollution to the environment. These elements are not only non-biodegradable and toxic but also accumulate in human organisms, causing some physical disorders and health concerns [4–6]. Therefore, WCS is classified as hazardous waste [7–9]. Unfortunately, more than 80% of them are directly dumped without treatment, which may cause potential pollution [10,11].

In order to comply with strict environmental regulations, numerous studies have been conducted to upgrade WCS, including building material [12]. However, this method usually has new contamination, such as the leachability of zinc, arsenic and antimony, and can only accommodate a very small fraction [13,14].

Regarding the recovery of valuable metals from WCS, previous work has mainly focused on recycling elements from WCS using beneficiation, leaching and pyrometallurgical processes. These methods mainly include flotation [15,16], acid leaching [17,18], roasting followed by water leaching [19], bio-leaching [20] and direct reduction–magnetic separation or smelting reduction [21–23]. Roy et al. [15] adopted flotation to recover copper sulfide from copper smelter slag. The results show that under the optimal conditions, the recovery rate of copper sulfide reached 84.82%. There is no doubt that flotation can recover copper sulfide to a certain extent. However, the copper oxide within it cannot be recovered. Muhammad et al. [18] employed sulfuric acid to extract metals from copper-rich converter slag. A copper precipitation rate of 90% can be obtained. At the same time, the iron extraction rate is only slightly higher than 60%. Mikoda et al. [20] used acidithiobacillus thiooxidans to leach valuable metals from copper metallurgical slag. The extraction rate of rare earth element (REE) and Co can reach more than 90%. However, the process duration is too long. Physical beneficiation and hydrometallurgical process can efficiently extract Cu, Pb, REE, etc. However, these processes cannot effectively recover iron as the main element in copper slag.

The pyrometallurgical process is generally used to recover iron from copper slag [24]. Smelting reduction is a deep reduction process; during this process, FeO is reduced to pig iron by a reductant at a high temperature above 1450 °C [1]. It can obtain a high iron recovery rate, but shortens the use time of the kiln. It is effective to recover iron from copper slag (CS) by direct reduction–magnetic separation. By reducing CS at 1200–1300 °C, fayalite and copper sulfide can be transformed into metallic iron and copper, and then Fe-Cu alloy phase can be formed [25]. Subsequently, a magnetic separation process is used to recover the Fe-Cu alloy. However, magnetic separation tailings have not been effectively used.

In this paper, a comprehensive process for recovering valuable metals from WCS is developed. The main steps of the proposed process include flotation to enrich Cu and direct reducing–magnetic separation to produce crude Fe-Cu alloy, and subsequent non-magnetic tailings to replace clinker to produce common Portland cement. Meanwhile, lead and zinc were recovered as dust through this process. No tailings were produced during the entire process, indicating that this is a clean and green process.

2. Materials and Methods

2.1. Materials

The WCS used in this study was supplied by Anhui Province, China. The chemical composition of WCS is shown in Table 1. The WCS contains 39.54% Fe, 0.81% Cu and 32.11% SiO_2. Meanwhile, the contents of Pb and Zn are 0.12% and 2.35%, respectively, which can be volatilized from the reduction process. As shown in Figure 1, the main mineral phases of copper slag are magnetite (Fe_3O_4) and fayalite (Fe_2SiO_4). The microstructure of the copper slag was observed by SEM-EDS, and the result is shown in Figure 2. It can be seen from Figure 2 that iron mainly exists in the form of fayalite (Fe_2SiO_4) and magnetite (Fe_3O_4), while copper mainly exists in the form of matte (Cu_2S). Meanwhile, iron minerals and copper minerals exist as independent particles, which facilitates the recovery of copper through the flotation process.

Table 1. Chemical composition of waste copper slag (wt.%).

TFe	FeO	SiO_2	CaO	MgO	Al_2O_3	Cu	Pb	Zn	S	P	LOI *
39.54	40.96	32.11	2.78	1.65	3.34	0.81	0.12	2.35	0.48	0.11	1.02

* Loss on ignition.

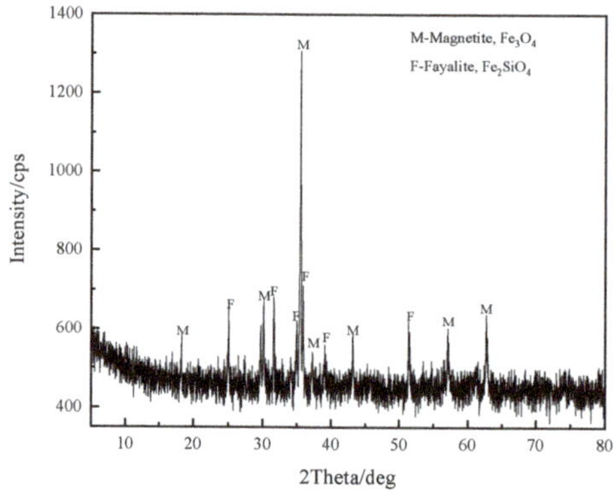

Figure 1. XRD pattern of waste copper slag.

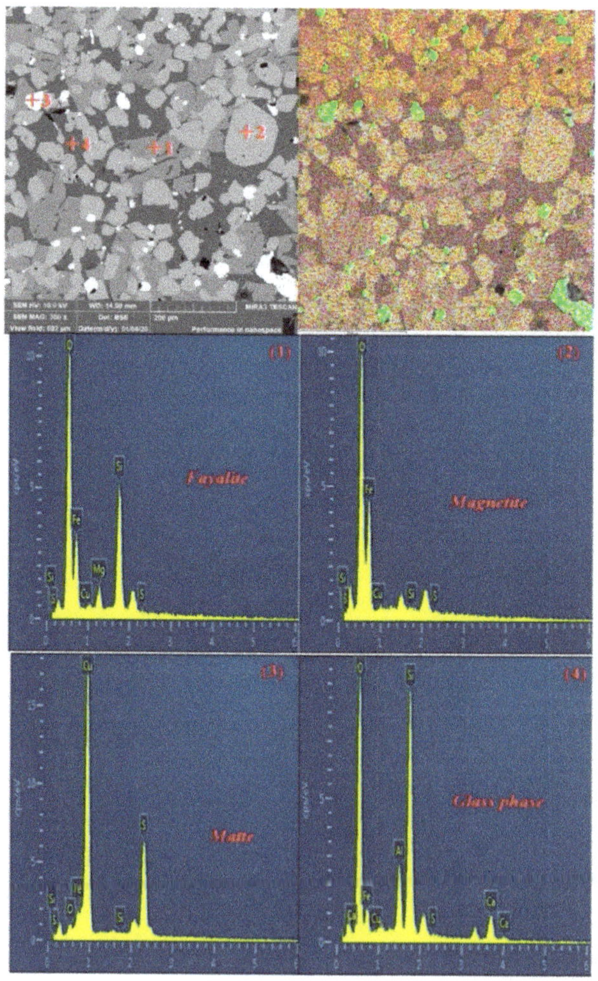

Figure 2. Microstructure of waste copper slag.(1, 2, 3, 4 represent mean points, element content in Table 2).

Table 2. Element content of Figure 2.

Point	Element/wt.%									Minerals
	Fe	Cu	S	Si	O	Mg	Al	Ca	Na	
1	49.15	0	0.25	16.18	31.55	2.87	0	0	0	Fayalite
2	70.32	0	0.05	0.59	29.04	0	0	0	0	Magnetite
3	11.81	62.92	22.31	0.07	2.88	0	0	0	0	Matte
4	14.14	0	0.38	29.34	42.61	0	6.35	7.18	0	Glass phase

2.2. Methods

The flow chart of the proposed WCS treatment process is presented in Figure 3, which was developed based on the basic understanding of the following steps: (1) the WCS is enriched with Cu by a flotation process; (2) direct reduction–magnetic separation of flotation tailing pellets to prepare Fe-Cu alloy and to remove Pb and Zn; (3) non-magnetic tailing to prepare Portland cement to achieve zero tailings. The experimental details of each step are described as follows.

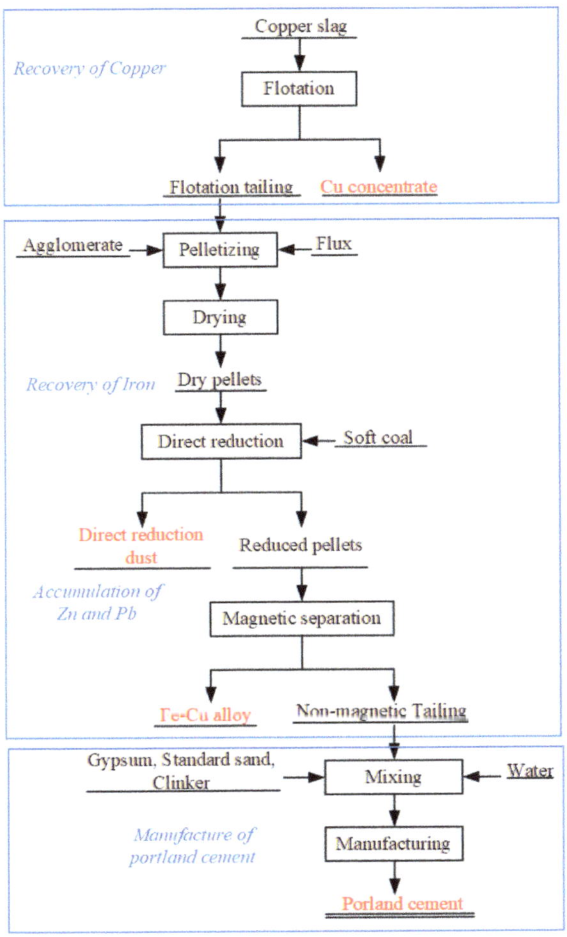

Figure 3. Experiment flow chart of the proposed process.

2.2.1. Flotation Process to Recover Cu

The flotation tests were carried out in a 0.75 L XFD laboratory flotation cell (FGC5-35, Yunhao Mining and Metallurgical Equipment Co., Ltd., Wuhan, China). In all flotation tests, a 500 g sample was used, the impeller speed was set to 600 rpm and the aeration rate was 0.2 dm^3/min. The specific operation of the flotation tests is outlined in Figure 4. In the flotation process, analytical grade Butyl xanthate ($C_5H_{10}OS_2$) (g/t) and terpene oil ($C_{10}H_{18}O$) (g/t) were used as collector and frother, respectively [7]. The recovery rate of copper is calculated as follows:

$$\eta_{Cu} = \frac{w_1}{w_0} \times y \times 100\% \quad (1)$$

where η_{Cu} is the recovery rate of copper; w_1 and w_0 are the copper grades in the flotation concentrate and raw material, respectively; y is the yield of the flotation concentrate.

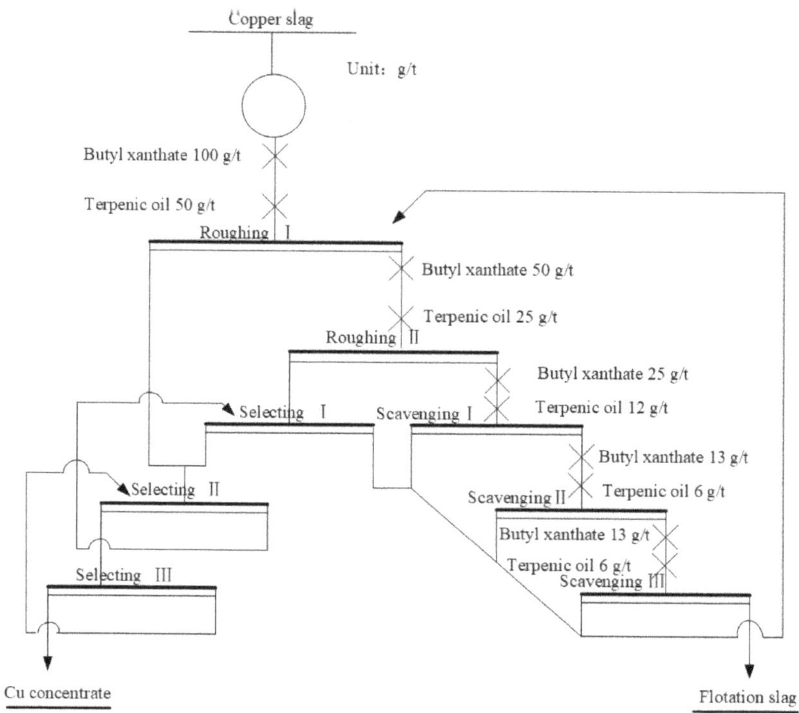

Figure 4. Schematic diagram of the flotation process.

2.2.2. Preparation of Crude Fe-Cu Alloy by Direct Reduction–Magnetic Separation

The flotation tailings and flux were fully mixed, and then the mixture was pressed into cylindrical briquettes (height: 10 mm and diameter: 10 mm). The briquettes were dried at 110 °C for 4 h.

The direct reduction tests were performed in a muffle furnace at 1250 °C for 80 min. For the reduction test, 60 g briquettes were mixed with 120 g soft coal (ash content: 4.49%; production location: Shenfu) (1–5 mm) and then encased into a corundum crucible with a depth of 80 mm and diameter of 80 mm. [21]. After reduction, the reduced briquettes were cooled to room temperature for the subsequent magnetic separation process. The specific operation process of magnetic separation is outlined in reference [26]. The metal recovery rate (iron or copper) is calculated as follows:

$$\eta = \frac{w'}{w} \times y_1 \times 100\% \tag{2}$$

where η is the metal recovery rate (iron or copper); w' and w are the iron (or copper) grades in the magnetic separation concentrate and reduced pellets, respectively; and y_1 is the yield of the concentrate.

The volatilization rate of zinc (or lead) is calculated as follows:

$$\gamma = \left(100 - \frac{w_2}{w_3} \times y_2\right) \times 100\% \tag{3}$$

where γ is the volatilization rate of zinc (or lead); w_2 and w_3 are the content of zinc (or lead) in the reduced pellets and flotation, respectively; and y_2 is the yield of reduced pellets.

2.2.3. Production of Common Portland Cement

A certain proportion of non-magnetic material (NMT), clinker and gypsum were thoroughly mixed. Ingredients of Portland cement are shown in Table 3.

Table 3. Ingredients of Portland cement/%.

Sample	Non-Magnetic Tailing	Gypsum	Clinker	Standard Sand/Cement Ratio	Water/Cement Ratio
PC0	0	5	95	9	0.5
PC6	6	5	89	9	0.5
PC9	9	5	86	9	0.5
PC12	12	5	83	9	0.5
PC15	15	5	80	9	0.5

The mixtures were cast into steel molds (cuboid) in three layers and compacted using a vibrating table [27]. All samples were stored in the ambient environment under a plastic sheet, and then the samples were demolded and wet-cured until the test date. The samples were cured at room temperature (18–22 °C) for 3 and 28 days separately. Concrete cuboids with a size of 40 × 40 × 160 mm were cast for the compressive strength test (Equipment: Universal testing machine, KY-D2305, Jilin Province Jilin Test Technology Co., Ltd., Jilin, China).

2.3. Characterization of Raw Materials and Products

The chemical compositions of WCS, flotation slag, Fe-Cu alloy and non-magnetical material tailings were measured by X-ray fluorescence spectroscopy (XRF, PANalytical Axios mAX, PANalytical B, V., Almelo, The Netherlands). The microstructure of reduced briquettes was observed and characterized by scanning electron microscopy (SEM) (TESCAN, MIRA3, LMH, FEI, Eindhoven, The Netherlands) and energy-dispersive spectrometry (EDS) (XMAX20, FEI, Eindhovern, The Netherlands). The mineral phases of WCS, flotation tailing, direct reduction dust and reduced briquettes were measured by X-ray powder diffraction (XRD, RIGAKU, D/Max-2500, Bruker corporation, Madison, WI, USA) with a 2θ scan range from 10° to 80°. The compressive strength of the Portland cement was measured according to the national standard (GB/T 175-2007).

3. Results and Discussion

3.1. Flotation Process to Enrich Cu in Waste Copper Slag

Table 4 lists the chemical composition of the flotation concentrates and tailings. When the overall copper recovery rate was 77.78%, the Cu content of the flotation concentrate was 21.50%. In addition, the contents of Fe_{total} and SiO_2 in the concentrate were higher, wth values are 28.22% and 21.33%, respectively. Meanwhile, the contents of S, Pb and Zn were 10.15%, 0.77% and 0.72%, respectively. The contents of Fe, SiO_2, Pb and Zn of the tailings were 39.88%, 32.44%, 0.1% and 2.4%, respectively. The size distribution of the Cu concentrate and flotation tailings is presented in Figure 5a,b. It can be seen that their particle size is fine, and 80% pass 100 μm.

Table 4. Chemical compositions of the flotation concentrate and tailings (wt.%).

Sample	TFe	SiO_2	CaO	MgO	Al_2O_3	Cu	Pb	Zn	S
Raw WCS	39.54	32.11	2.78	1.65	3.34	0.81	0.12	2.35	0.48
Concentrate	28.22	21.33	2.34	1.02	1.71	21.50	0.77	0.72	10.15
Tailing	39.88	32.44	2.79	1.67	3.39	0.25	0.1	2.4	0.19

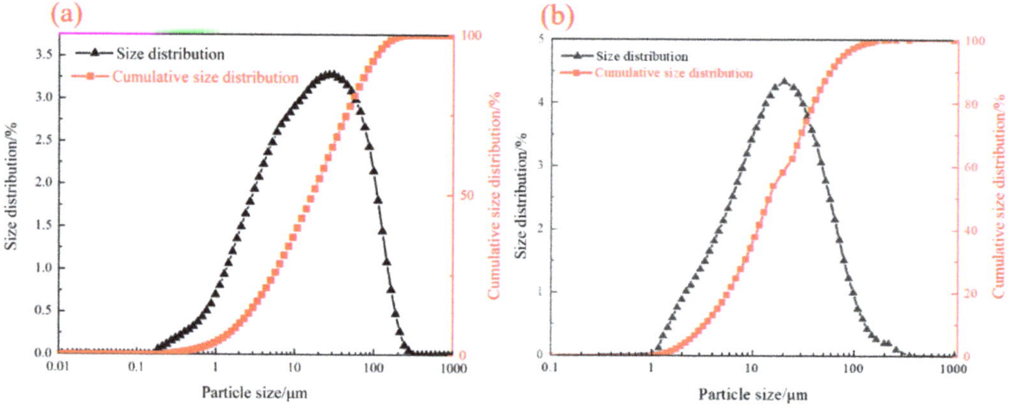

Figure 5. Size distributions of Cu concentrate and flotation tailings (**a**) Cu concentrate, (**b**) flotation tailings.

The XRD patterns of the Cu concentrate and tailings are shown in Figure 6a,b. It indicates that iron in the flotation tailings mainly exists in the form of fayalite (Fe_2SiO_4) and magnetite (Fe_3O_4). The microstructure of copper concentrate was observed by SEM-EDS, and the results are shown in Figure 7a–d. It can be seen from Figure 7 that chalcopyrite ($CuFeS_2$) and copper sulfide (Cu_2S) are the main mineral phases of copper-bearing with a particle size of less than 50 μm.

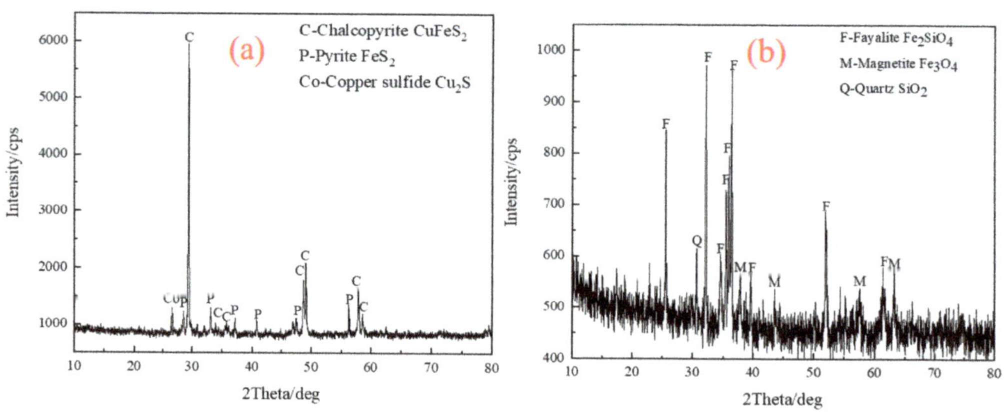

Figure 6. XRD of Cu concentrate and flotation tailings (**a**) Cu concentrate, (**b**) flotation tailings.

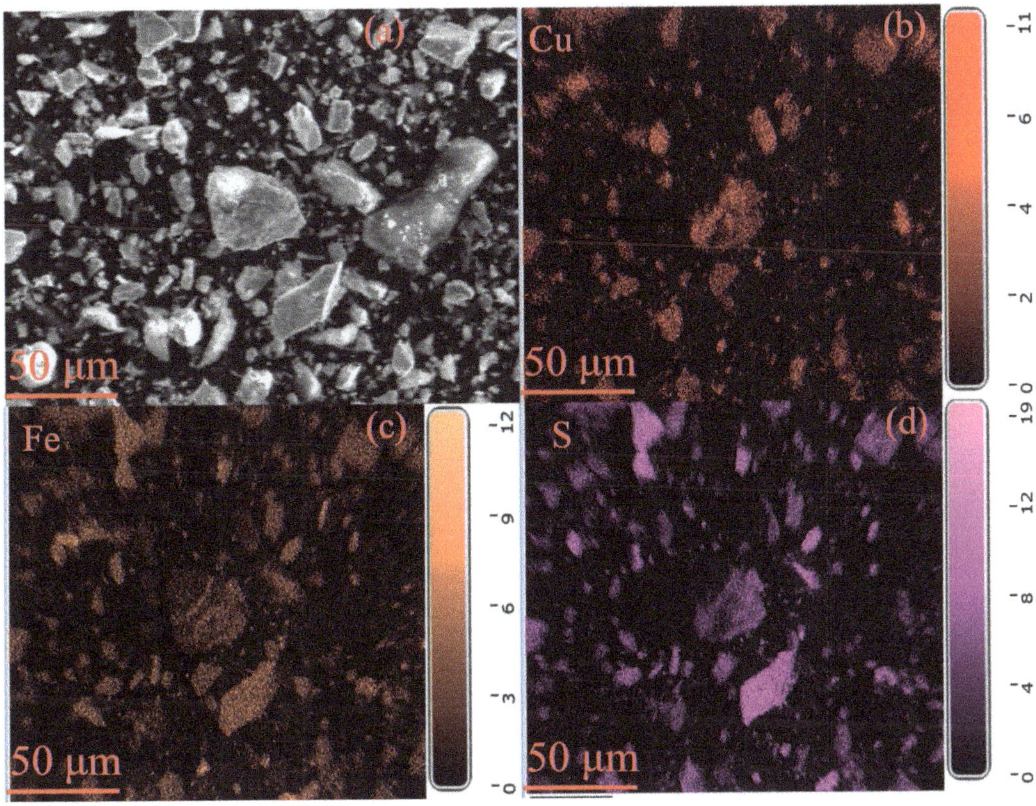

Figure 7. SEM-EDS of Cu concentrate (**a**) is the original SEM, (**b–d**) are the mapping scanning of copper, iron and sulfur, respectively).

3.2. Direct Reduction–Magnetic Process to Recover Fe, Cu, Pb and Zn

According to Figure 6, the main phases of flotation tailings are fayalite and magnetite. The reduction of magnetite, fayalite and pyrite has been investigated in previous research [10,21,23]. These studies have shown that it is difficult to reduce fayalite with CO and that it can only be reduced by carbon at high temperatures. In the presence of CaO, fayalite is more easily reduced by C and CO. Pyrite and copper sulfide are difficult to be reduced by solid carbon to form metallic iron and copper. Adding CaO can reduce the required Gibbs free energy of pyrite and copper sulfide reduction.

Figure 8 presents the effect of additive dosage on the recovery of Fe and Cu and the volatilization of Zn and Pb. Without additives, the recovery rates of iron and copper are only 72.87% and 65.34%, respectively. Meanwhile, the iron grade of magnetic concentrate is 78.62%. Compared with no additives, in the presence of 20% additive, the recovery rates of iron and copper are 91.34% and 83.41%, respectively, and the grade of iron is 90.21%, and the content of copper is maintained at 0.34%. When the additives are 20% and 0%, the volatilization rates of zinc and lead are 98.89% and 95.33%, and 89.6% and 86.33%, respectively. Increasing the dosage to 25% has a slight effect on the indexes. From the above, it is recommended to add 20% as the best additive dosage.

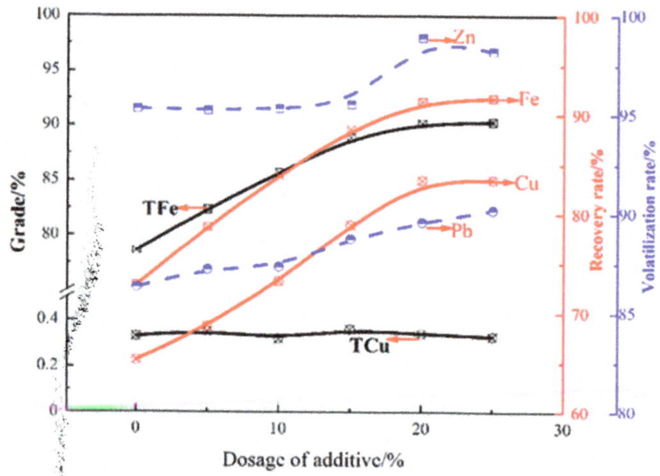

Figure 8. Effect of additive dosage on recovery of Fe, Cu, Zn and Pb (reduction at 1250 °C for 80 min at C/Fe mass ratio of 2, 0.3 basicity).

The XRD patterns of the roasted briquettes obtained under different additive dosages are presented in Figure 9. It can be seen from Figure 9 that during the reduction process, the peak intensity of iron is enhanced in the presence of additives, which is caused by the iron grain growing. Figure 10 shows the electron probe microanalysis (EPMA) of roasted briquettes with 20% additives. The roasted briquettes contain two parts,:one part is Fe-Cu alloy, and the other part is gangue. Most of the copper is combined with iron to form Fe-Cu alloy. Meanwhile, the boundary between the alloy and the gangue is obvious, which facilitates the recovery of iron and copper through the grinding process.

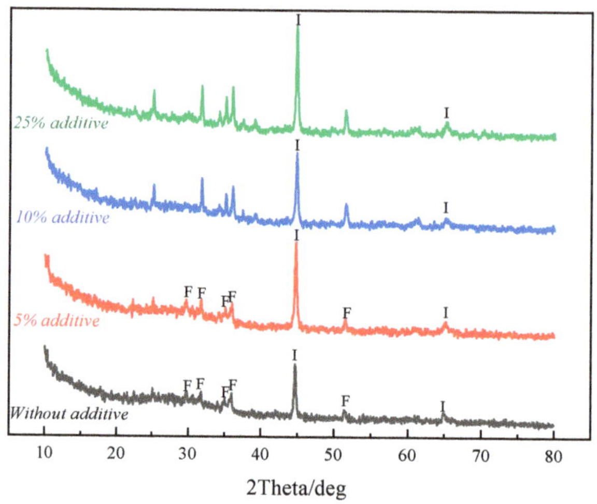

Figure 9. XRD patterns of reduced pellets obtained at the different additive dosages.

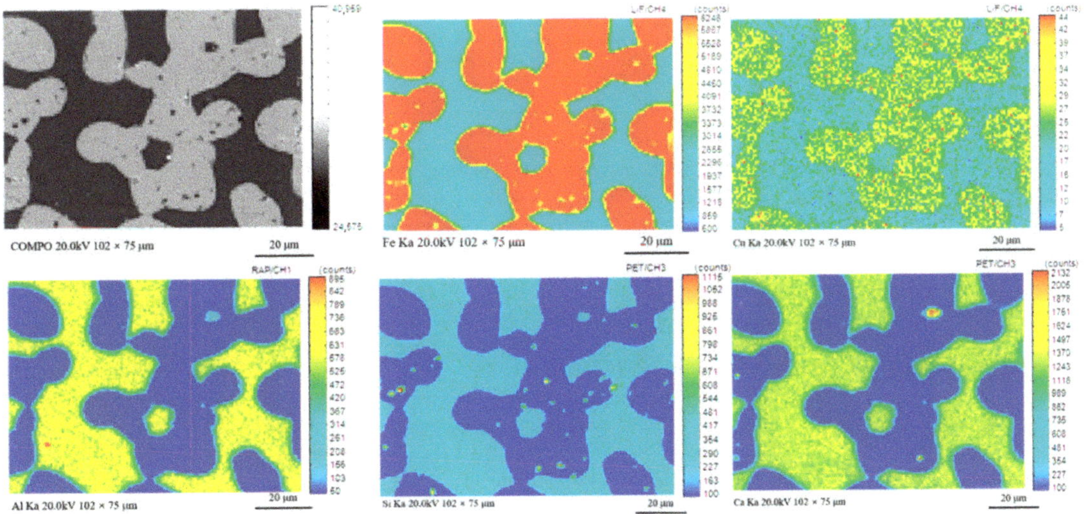

Figure 10. Electron probe microanalysis (EPMA) of reduced pellets with 20% additives (F: Fayalite Fe_2SiO_4; I: Iron Fe).

Reduction was performed at 1250 °C for 80 min under the conditions of C/Fe mass ratio of 2, the basicity of 0.3 and additive dosages of 20%. The chemical compositions of the Fe-Cu alloy obtained at 20% additive are presented in Table 5. It can be seen that the Fe-Cu alloy contains 90.21% of TFe and 0.4% of Cu, and the contents of S and P are only 0.032% and 0.025%, respectively, which can be used as a burden for the weather resistance steelmaking. The results further confirm that the direct reduction–magnetic separation process is an effective means to upgrade the WCS.

Table 5. Chemical compositions of Fe-Cu alloy obtained with 20% additives.

Sample	TFe	SiO_2	CaO	MgO	Al_2O_3	Cu	Pb	Zn	S	P
Fe-Cu alloy	90.21	3.32	0.89	0.24	0.32	0.4	0.044	0.021	0.032	0.025

The chemical compositions of the direct reduction dust are shown in Table 6. The main compositions of dust are 65.17% ZnO and 2.66% PbO, which can be used as raw materials for extracting zinc. Many studies have been carried out on the comprehensive utilization of secondary zinc oxide [28,29]. Therefore, the treatment of reduction dust are not discussed in this paper.

Table 6. Chemical compositions of dust/wt.%.

Sample	ZnO	PbO	C	SiO_2	CaO	MgO	Al_2O_3	Na_2O	S
Dust	65.17	2.66	8.99	1.78	0.09	0.14	0.22	0.56	2.19

3.3. *Preparation of Portland Cement from Non-Magnetic Tailings*

The chemical composition of non-magnetic tailings is listed in Table 7. The non-magnetic tailings include 55.64% SiO_2 and 16.83% CaO, and the contents of Cu, Pb and Zn are 0.01%, 0.06% and 0.02%, respectively. According to the USEPA standard, the toxicity characteristics of the non-magnetic tailings were determined by the leaching procedure tests, and the results are shown in Table 8. The concentrations of Cu^{2+}, Pb^{2+} and Zn^{2+} in the leachate are 0.021 mg/L, 0.001 mg/L and 0.046 mg/L, respectively, which are lower than the critical value, indicating that the non-magnetic tailings harmless and can be used directly as the burden for preparing cement.

Table 7. Chemical compositions of non-magnetic tailings/wt.%.

Sample	TFe	SiO_2	CaO	MgO	Al_2O_3	Cu	Pb	Zn	S	P
Tailing	9.62	55.64	16.83	2.91	5.99	0.01	0.06	0.02	0.15	0.18

Table 8. Toxicity Characteristics of Leaching Procedure (TCLP) test results of non-magnetic tailings/mg/L.

Elements	Cu	Pb	Zn
Tailing	0.021	0.001	0.046
Toxicity thresholds	100	5	100

Figure 11 and Table 9 present the particle size distributions of non-magnetic tailings obtained from different ball-milling times. It can be seen that after ball milling the original sample for 1 h, 90% of the powder is less than 30 μm. When ball milling was extended to 3 h, the particle size changed slightly. When the ball milling time was further extended to 4 h, the powder agglomerated, and the particle size D90 increased. The specific surface area (SSA) increased from 1.03 g/cm^2 to 2.73 g/cm^2. Considering the cost, it is recommended to use a ball milling of 2 h. The samples obtained from the ball milling for 2 h were applied as raw materials for cement manufacturing.

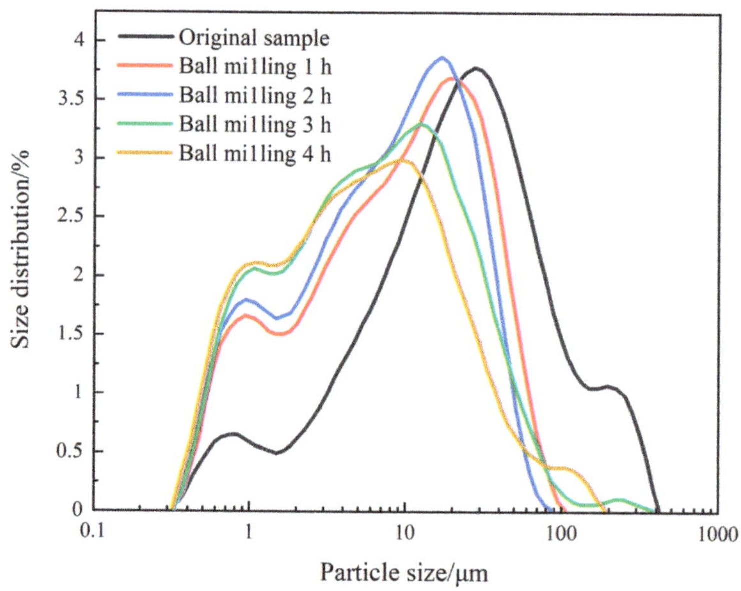

Figure 11. Particle size distributions of non-magnetic tailings obtained from different ball milling times.

Table 9. Particle size parameters of non-magnetic tailings obtained from different ball milling times.

Particle Size/μm	Original Sample	Ball Milling 1 h	Ball Milling 2 h	Ball Milling 3 h	Ball Milling 4 h
D10	2.66	0.87	0.85	0.78	0.75
D50	21	7.67	6.52	5.22	5.10
D90	106.5	29.68	24.05	23.58	29.7
SSA/g/cm^2	1.03	2.22	2.34	2.63	2.73

According to the Chinese Standard GB/T 175-2007 (Standard for Common Portland cement), the minimum compressive strength of Portland cement at 3 and 28 days should be greater than 17 and 42.5 MPa, respectively (GB/T 175-2007). They explored the most suitable mixing ratio of copper slag tailings to replace clinker. They believed that the proportion of copper slag tailings ranges from 6% to 15%.

Figure 12 shows the changes of the compressive strength and breaking strength versus the clinker content replacement by non-magnetic tailings at 3 and 28 days. It can be seen that the compressive strength of the Portland cement at 3 and 28 days increased from 24.1 to 29.70 MPa and 47 to 50.89 MPa, respectively, and the addition of non-magnetic tailings increased from 0% to 9%. If the amount of non-magnetic tailings is further increased, the compressive strength is decreased. Compared with the 42.5 MPa compressive strength level of common Portland cement strength standard, the compressive strength of the samples at 3 and 28 days are higher. Meanwhile, with the increase in the amount of non-magnetic tailings, the breaking strength at 3 and 28 days changed slightly, and their values were about 5 and 7.5 MPa, respectively. All samples meet the 42.5 strength level of common Portland cement strength standard.

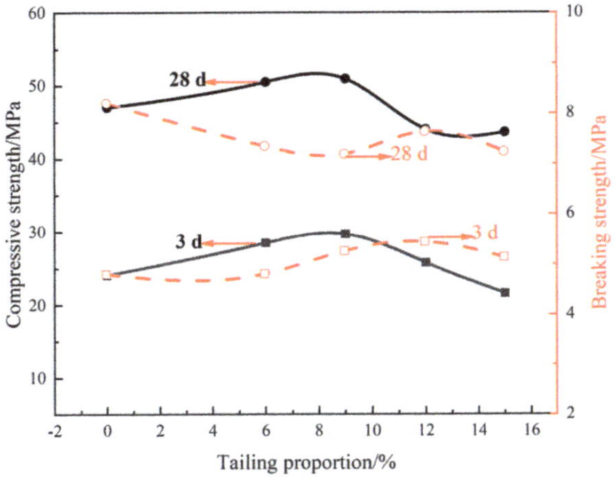

Figure 12. Variations of compressive strength versus the content of cement replacement by non-magnetic tailings.

3.4. Balance of Elements in the Proposed Flow Chart

From the above research, we proposed a stepwise extraction process for the recovery of Cu, Fe, Zn and Pb, and the production of common Portland cement from non-magnetic tailings, as shown in Figure 13. It can be seen that the reason for the increase in the yield of green pellets is the addition of limestone and binder. However, through direct reduction, the sum of dust yield and reduced pellet yield is less than the yield of green pellets. This is because of the loss on ignition of limestone and flotation tailings. Several products were obtained from WCS, Cu concentrate for copper smelting, magnetic concentrate for weather resistance steelmaking and non-magnetic tailings for building construction. The proposed process can recover more than 90% iron and copper. Meanwhile, the recovery rates of Zn and Pb exceed over 90%. Moreover, the proposed process achieves zero tailing emissions. Therefore, WCS can be effectively utilized through this process.

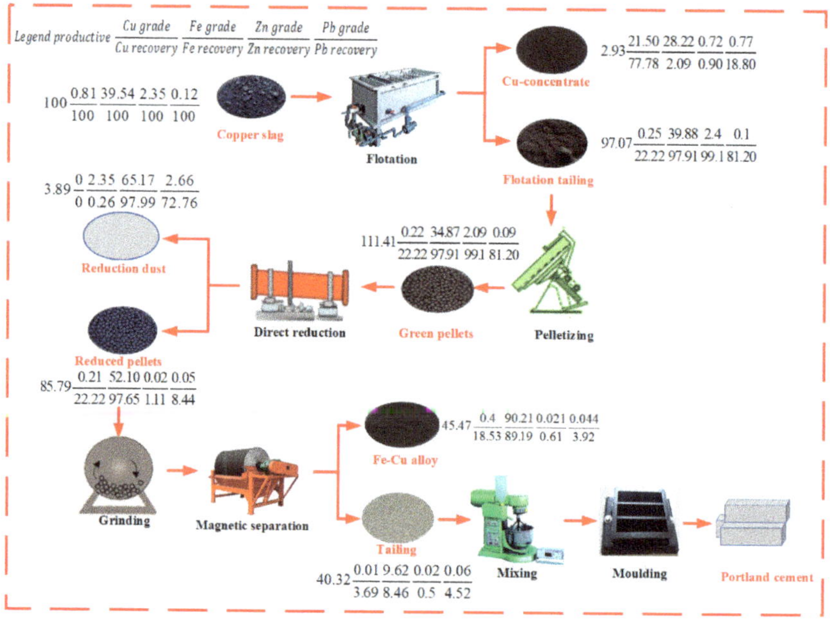

Figure 13. Balance of elements in the proposed flowchart.

4. Conclusions

The waste copper slag containing 39.54% Fe and 0.81% Cu was used as a raw material for recovering valuable elements through stepwise extraction. Under the optimal conditions, four products were obtained, including flotation concentrate, magnetic separation concentrate, zinc-bearing dust and Portland cement. The overall Cy recovery rate of the flotation concentrate is 77.78%, which can reach a copper content of 21.50%, which can be used as a burden for copper smelting. The magnetite concentrate contains 90.21% of Fe and 0.4% of Cu with 91.34% of iron recovery and 83.14% of copper recovery, which meets the requirements of weather steelmaking. Meanwhile, the zinc-bearing dust contains 65.17% of ZnO and 2.66% of PbO with an overall zinc and lead volatilization of 98.89% and 89.6%, which can be used as a raw material for preparing nano ZnO. The strength of common Portland cement meets the 42.5 strength level standard. The proposed process can provide an alternative method for the effective and green utilization of waste copper slag.

Author Contributions: Methodology, Z.G.; Investigation, S.L. (Siwei Li), T.D. and S.L. (Shenghu Lu); Writing—original draft preparation, S.L. (Siwei Li); Writing—review and editing, S.L. (Siwei Li); Supervision, J.P. and D.Z.; Funding acquisition, Z.G. and D.Z. All authors have read and agreed to the published version of the manuscript.

Funding: This work was financially supported by the Youth natural science foundation China (No. 51904347), Innovation-driven Project of Guangxi Zhuang Autonomous Region (No. AA18242003, No. AA148242003) and the Youth Natural Science Foundation of Hunan Province (No. 2020JJ5726).

Data Availability Statement: Not applicable.

Acknowledgments: The authors would like to thank Co-Innovation Center for Clean and Efficient Utilization of Strategic Metal Mineral Resources of Hunan Province, which supplied us with the facilities and funds to fulfill the experiments. The authors would like to express their gratitude to EditSprings (http://www.editsprings.com) for the expert linguistic services provided.

Conflicts of Interest: The authors declare no conflict of interest.

References

1. Sarfo, P.; Das, A.; Wyss, G.; Young, C. Recovery of metal values from copper slag and reuse of residual secondary slag. *Waste Manag.* **2017**, *70*, 272–281. [CrossRef]
2. Shen, H.; Forssberg, E. An overview of recovery of metals from slags. *Waste Manag.* **2003**, *23*, 933–949. [CrossRef]
3. Heo, J.H.; Kim, B.S.; Park, J.H. Effect of CaO addition on iron recovery from copper smelting slags by solid carbon. *Metall. Mater. Trans. B* **2013**, *44*, 1352–1363. [CrossRef]
4. Heo, J.H.; Chung, Y.S.; Park, J.H. Recovery of iron and removal of hazardous elements from waste copper slag via a novel aluminothermic smelting reduction (ASR) process. *J. Clean. Prod.* **2016**, *137*, 777–787. [CrossRef]
5. Karwowska, E.; Wojtkowska, M.; Andrzejewska, D. The influence of metal speciation in combustion waste on the efficiency of Cu, Pb, Zn, Cd, Ni and Cr bioleaching in a mixed culture of sulfur-oxidizing and biosurfactant-producing bacteria. *J. Hazard Mater.* **2015**, *299*, 35–41. [CrossRef]
6. Saffarzadeh, A.; Shimaoka, T.; Motomura, Y.; Watanabe, K. Characterization study of heavy metal-bearing phases in MSW slag. *J. Hazard Mater.* **2009**, *164*, 829–834. [CrossRef]
7. Alp, I.; Deveci, H.; Süngün, H. Utilization of flotation wastes of copper slag as raw material in cement production. *J. Hazard Mater.* **2008**, *159*, 390–395. [CrossRef]
8. Durinck, D.; Engstrom, F.; Arnout, S.; Heulens, J.; Jones, P.T.; Bj€orkman, B.; Blanpain, B.; Wollants, P. Hot stage processing of metallurgical slags. *Resour. Conserv. Recycl.* **2008**, *52*, 1121–1131. [CrossRef]
9. Shi, C.; Meyer, C.; Behnood, A. Utilization of copper slag in cement and concrete. *Resour. Conserv. Recycl.* **2008**, *52*, 1115–1120. [CrossRef]
10. Guo, Z.Q.; Zhu, D.Q.; Pan, J.; Zhang, F. Innovative methodology for comprehensive and harmless utilization of waste copper slag via selective reduction-magnetic separation process. *J. Clean. Prod.* **2018**, *187*, 910–922. [CrossRef]
11. Chen, M.; Zeng, G.; Huang, D.; Yang, C.; Lai, C.; Liu, Y.; Xu, P.; Zhang, C.; Wang, J.; Liu, Y.; et al. Degradation of atrazine by a novel Fenton-like process and assessment the influence on the treated soil. *J. Hazard Mater.* **2016**, *312*, 184–191. [CrossRef] [PubMed]
12. Kalusuraman, G.; Kumaran, S.; Aslan, M.; Küçükömerog˘luc, T.; Siva, I. Use of waste copper slag filled jute fiber reinforced composites for effective erosion prevention. *Measurement* **2019**, *148*, 106950. [CrossRef]
13. Potysz, A.; Hullebusch, E.D.; Kierczak, J.; Grybos, M.; Lens, P.N.; Guibaud, G. Copper metallurgical slags—Current knowledge and fate: A review. *Crit. Rev. Environ. Sci. Technol.* **2015**, *45*, 2424–2488. [CrossRef]
14. Murari, K.; Siddique, R.; Jain, K. Use of waste copper slag, a sustainable material. *J. Mater. Cycles Waste Manag.* **2015**, *17*, 13–26. [CrossRef]
15. Roy, S.; Datta, A.; Rehani, S. Flotation of copper sulphide from copper smelter slag using multiple collectors and their mixtures. *Int. J. Miner. Process.* **2015**, *143*, 43–49. [CrossRef]
16. Guo, Z.Q.; Zhu, D.Q.; Pan, J.; Wu, T.J.; Zhang, F. Improving beneficiation of copper and iron from copper slag by modifying the molten copper slag. *Metals* **2016**, *6*, 86. [CrossRef]
17. Gargul, K.; Boryczko, B.; Bukowska, A.; Jarosz, P.; Małecki, S. Leaching of lead and copper from flash smelting slag by citric acid. *Arch. Civ. Mech. Eng.* **2019**, *19*, 648–656. [CrossRef]
18. Muhammad, K.K.; Joseph, H.; Vivek, A.; Jouni, P.; Mika, H.; Mari, L. Sulfuric acid leaching for capturing value from copper rich converter slag. *J. Clean. Prod.* **2019**, *25*, 1005–1013.
19. Nadirov, R.K.; Syzdykova, L.I.; Zhussupova, A.K.; Usserbaev, M.T. Recovery of value metals from copper smelter slag by ammonium chloride treatment. *Int. J. Miner. Process* **2013**, *124*, 145–149. [CrossRef]
20. Mikoda, B.; Potysz, A.; Kmiecik, E. Bacterial leaching of critical metal values from polish copper metallurgical slags using acidithiobacillus thiooxidans. *J. Environ. Manag.* **2019**, *236*, 436–445. [CrossRef]
21. Li, S.W.; Pan, J.; Zhu, D.Q.; Guo, Z.Q.; Xu, J.W.; Chou, J.L. A novel process to upgrade the copper slag by direct reduction-magnetic separation with the addition of Na_2CO_3 and CaO. *Powder Technol.* **2019**, *347*, 159–169. [CrossRef]
22. Zhou, S.W.; Wei, Y.G.; Li, B.; Wang, H. Cleaner recycling of iron from waste copper slag by using walnut shell char as green reductant. *J. Clean. Prod.* **2019**, *217*, 423–431. [CrossRef]
23. Guo, Z.Q.; Pan, J.; Zhu, D.Q.; Zhang, F. Green and efficient utilization of waste ferric-oxide desulfurizer to clean waste copper slag by the smelting reduction-sulfurizing process. *J. Clean. Prod.* **2018**, *199*, 891–899. [CrossRef]
24. Keiran, H.; Hürman, E.R.; Pekka, T.; Ari, J. Upgrading copper slag cleaning tailings for re-use. *Miner. Eng.* **2019**, *133*, 35–42.
25. Zhou, X.L.; Zhu, D.Q.; Pan, J.; Wu, T.J. Utilization of waste copper slag to produce directly reduced iron for weathering resistant steel. *ISIJ* **2015**, *55*, 1347–1352.
26. Zhu, D.Q.; Xu, J.W.; Guo, Z.Q.; Pan, J.; Li, S.W.; Pan, L.T.; Yang, C.C. Synergetic utilization of copper slag and ferruginous manganese ore via co-reduction followed by magnetic separation process. *J. Clean. Prod.* **2020**, *250*, 119462. [CrossRef]

27. Shaker, F.; Rashad, A.; Allam, M. Properties of concrete incorporating locally produced Portland limestone cement. *Ain Shams Eng. J.* **2018**, *9*, 2301–2309. [CrossRef]
28. Huang, Y.K.; Geng, Y.B.; Han, G.H.; Cao, Y.J.; Peng, W.J.; Zhu, X.F.; Zhang, T.A.; Dou, Z.H. A perspective of stepwise utilization of hazardous zinc plant purification residue based in selective alkaline leaching of zinc. *J. Hazard Mater.* **2020**, *389*, 122090. [CrossRef]
29. Zhang, H.X.; Yang, Y.T.; Liu, X.J. Study of cadmium-doped zinc oxide nanocrystals with composition and size dependent band gaps. *Chin. J. Chem. Phys.* **2018**, *31*, 197–202. [CrossRef]

minerals

MDPI

Article

Prediction of Postflotation Tailings Behavior in a Large Storage Facility

Magdalena Wróżyńska

Department of Construction and Geoengineering, Poznań University of Life Sciences, Wojska Polskiego 28, 60-637 Poznań, Poland; magdalena.wrozynska@up.poznan.pl

Abstract: Extracting and copper production on a large scale generates large volumes of postflotation mine tailings. The scale of operation and development of tailings storage facilities (TSFs) forces the use of innovative solutions enabling safe storage now and in the future. Any changes to the operation require multi-directional monitoring of the impact of these changes on storage safety. The ongoing exploitation will be ensured by expansion of the TSF and a change in tailings storage technology. This approach will preclude the need for changes to the new location, such as changes of land use, and will minimise the volume of mine waste. The paper presents the results of pilot studies carried out to implement the change in postflotation tailings storage technology at Żelazny Most TSF (Poland) in the future. The aim of the paper was settlements prediction of tailings and comparison of deformations with observed settlements. Settlements prediction of tailings was made on the basis of the results of the DMT (Marchetti Dilatometer Test), recommended for the prediction of natural soil settlement. Depending on the analysed zone of the TSF, settlements ranged from a few centimetres to over 1.5 m. Despite the difference shown, the results of DMT and geodetic measurements indicate a convergent trend of settlement.

Keywords: tailings storage facility; copper tailings settlement; experimental embankment; Marchetti Dilatometer Test; geodetic measurements

Citation: Wróżyńska, M. Prediction of Postflotation Tailings Behavior in a Large Storage Facility. *Minerals* **2021**, *11*, 362. https://doi.org/10.3390/min11040362

Academic Editors: Chongchong Qi, Guichen Li, Hakan Basarir, Qiusong Chen and Yuantian Sun

Received: 3 March 2021
Accepted: 28 March 2021
Published: 30 March 2021

1. Introduction

The tailings storage facility (TSF) is one of the key links in the production of copper concentrate, without which manufacturing raw material would be impossible. The issues of the currently exploited TSF are extensive and complicated, from engineering problems and challenges, through economic to environmental and social aspects [1–4]. In Poland, since 1977, there has been only one storage facility for tailings from copper ore flotation for all three active mines of KGHM Polska Miedź SA (Polish Copper). The feature of such a large tailings storage facility, which enables comprehensive waste management, is the low efficiency of utilizing the facility's capacity in relation to the amount of deposited waste, which currently accounts for about 95% of the excavated mining waste. The reason for the unfavourable balance is the low density of the material deposited in the form of sludge, ranging from 1.11 and 1.15 Mg/m^3. Further deposition with such large discharges of waste will lead to exhaustion of the operational possibilities of the storage facility. One of the solutions to this problem may be the construction of another tailings storage facility. Another more rational solution seems to be the implementation of a more effective method of depositing waste, e.g., pre-thickened paste. A third approach combining the two previous solutions can also be considered. As far as this solution is concerned, the simultaneous expansion of the existing storage facility and change in the technology used for depositing postflotation tailings would exclude the need to build another facility in a new location. This would also increase the operational capability of the currently used facility and would extend its operation time. However, the implementation of the third solution requires prior assessment of the suitability of the deposited tailings as a subsoil on which higher density waste (paste) will be stored. Such an analysis is indispensable in predicting the behaviour of the massif of tailings loaded with paste and can be additionally used to assess the overall stability of the entire tailings storage facility.

An experimental embankment was built in order to determine the real reaction of the subsoil made of tailings to the additional load. However, due to the fact that the storage facility is an earth structure, which is diverse and spatially variable in terms of grain size and physical and mechanical properties, the settlement process in these conditions depends on many local factors. In order to learn about the impact of these factors, a number of studies and observations of loaded tailings were carried out as part of a three-year monitoring. Geotechnical monitoring consisted of a series of in situ tests and laboratory tests of extracted tailings samples, among others Marchetti Dilatometer Tests (DMTs), cone penetration tests (CPTUs), and field vane tests (FVTs). Measurements of real displacements covered by geodetic monitoring were also carried out throughout the observation three-year period.

Many researchers and world experts attempted to predict natural soil settlement based on the results of CPTUs and DMTs, indicating both the advantages and disadvantages of these tests [5]. Predicting the settlement of natural soil on the basis of CPTUs is well recognized and widely used [6–16]. In order to estimate the settlement of natural soil based on CPTUs, a constrained modulus M_{CPTU} needs to be determined [11,17–23]. The research conducted by Robertson et al. [24] proves that the estimation of M_{CPTU} only on the basis of the cone resistance value may be inaccurate and even burdened with a large error, especially in the case of pre-consolidated soils. The collapse mechanism under the cone, resulting in large soil deformations during the test, poorly corresponds to M_{CPTU}. According to Jamiolkowski et al. [25], it is impossible to estimate the exact value of M_{CPTU} only on the basis of cone resistance without knowing the stress history. The selection of a reliable empirical factor α, a parameter necessary to calculate M_{CPTU} based on the cone resistance [26], may also pose a difficulty. Due to the abovementioned limitations, CPTU results should be calibrated based on the results of laboratory tests or other in situ tests [27]. Greater accuracy of estimated settlements in relation to the possibilities offered by CPTU can be obtained from the DMT [28,29]. The advantage of the DMT over CPTU that is commonly emphasized in the literature is its advantage in assessing soil deformation parameters [30]. This is due to the test conditions, where in the CPTU the measurement performed with the cone is related to the bearing capacity of the subsoil, while in the DMT, the dilatometer measurement is related to the subsoil deformability. Available research results from many authors clearly indicate that the DMT is meaningful and reliable in estimating natural soil settlement [31–50]. In order to estimate natural soil settlement based on DMT, the constrained modulus M_{DMT} needs to be determined. The authors of [35] recommend the estimation of settlements based on DMT in accordance with the classic uniaxial deformation method, analogously to the CPTU method. According to Marchetti [34], the results of settlements estimated using the 1-D elasticity formula do not significantly differ from the results estimated with the 3-D formula (triaxial deformations) and differences do not exceed 10%. From an engineering point of view, the 1-D model seems less problematic and more practical in this situation (e.g., it does not require the knowledge of Poisson's ratio ν). Determining and comparing settlements predicted with DMT with real observations (geodetic measurements), possibly with the results of tests other than DMT, was implemented by many authors. The high compatibility of the settlements predicted by DMT with the observed settlements for natural soils of different grain sizes was confirmed by [32,51]. The authors of [33,52] also recommend a procedure for estimating settlements based on DMT results. According to Monaco et al. [46], the Marchetti flat dilatometer is currently the most practical device for settlements prediction.

In terms of the purpose of the work indicating the possibility of continuing to exploit the TSF without having to build a new facility while not making changes to the new location (e.g., land use and land management change), the aims of the present research were settlements prediction (1) of tailings loaded with an experimental embankment using DMT, used for settlements prediction of natural soils; determination of observed settlements (2) based on the results of geodetic measurements; and, additionally, comparison of (3) predicted and observed settlements of tailings. In addition, the validity of using the DMT recommended for testing natural soils for general trend estimation of tailings settlement was confirmed.

2. Materials and Methods

2.1. Study Site

KGHM Polska Miedź SA is the main domestic copper manufacturer in Poland. During over 50 years of operation, over 18 million tons of copper were produced from extracted rock output, with a mass of 1 billion tons. Based on the estimated recognition of the deposit resources, copper ore can be exploited for the next 30 years. For annual production of approximately 2 million tons of concentrate with an average content of 23% of copper, from 4% to 6% of the extracted mining mass is utilised. The remaining 94–96% are postflotation tailings that require safe management. An important problem arising from such a significant scale of production of the raw materials, whose average content is 1.52% of copper in the deposits exploited in Poland, is the management of huge masses of postflotation tailings. These tailings are generated as a result of the following operations and processes: sifting, grinding, classification, flotation, densification, filtration, and drying. As a result of flotation, about 25 million Mg of waste is transported by the hydraulic method to the storage facility each year and it is rising. At present, the only active receiver of all postflotation tailings from Legnica-Głogów Copper District is Żelazny Most Tailings Storage Facility (Figure 1).

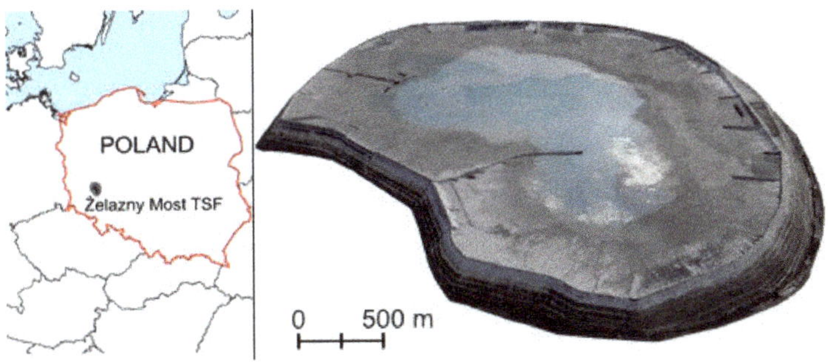

Figure 1. Location (**left**) and 3-D visualisation (**right**) of Żelazny Most tailings storage facility (TSF), Poland.

The storage facility is located in the Lower Silesia Voivodeship, the south-western part of Poland; a natural land depression in the south-eastern part of the Dalkowskie Hills, which is part of the Trzebnickie Hills, was used for its positioning. This depression is in the form of a natural kettle hole surrounded by frontal moraine-dammed hills of the Riss glaciation [4]. Żelazny Most TSF is an earth above-ground hydrotechnical construction, currently the largest in Europe and second largest in the world [2,53]. The construction of TSF took place between 1974 and 1977. Since the commissioning on 12 February 1977, the storage facility has been subjected to continuous and simultaneous operation and expansion, because of which it is described as "atypical" [3]. Superstructuring of the facility's dams is carried out with the upstream method. The facility's extension module is 2.5 m, which results in an average annual increase in dam height of about 1.3 m [54]. For the continuous superstructuring of dams surrounding the facility, waste of the coarse grains, previously deposited within the so-called beaches, is used. Due to favourable strength and filtration characteristics, meeting the criteria of grain size and ultimately density, sandy postflotation tailings is a full-fledged building material in this place [55,56]. TSF is currently in the process of forming dams up to 190 m above sea level, which results in the possibility of storing postflotation tailings of a volume of 780 million m^3. At the current stage of TSF expansion, one of the important factors limiting its further development in the technology implemented so far is the complex geological structure of the natural subsoil of the facility. The superstructure of the existing dams to the crown's ordinate of 195 m above sea level along with the extension of the facility by an additional effective area, the so-called "Southern Section" with an area of over 600 ha, would allow for the deposit of 950 million m^3 of postflotation tailings. In the long run, this provides the possibility of storing tailings until 2037. Operation of the facility after expansion can be continued using the existing deposition method ("wet deposition") or the mixed method. The latter combines the wet deposition method with the method of depositing tailings densified to the consistency of paste. The application of the paste deposition method requires the design and construction of technical infrastructure for densifying tailings. The parallel method may be to collect dehydrated and pre-thickened tailings in a natural way that was previously deposited inside the storage facility. For the first time, an industrial depositing method was used at the Kidd Creek mine in Canada in 1973, although research had already been carried out earlier [57,58]. This method can definitely compete with the currently used method of depositing in Żelazny Most TSF. The implementation of a new method of depositing tailings masses of higher density than before on a highly deformable subsoil must be preceded by a series of tests and observations of the behaviour of TSF under the influence of an additional load.

2.2. Experimental Embankment Characteristics

Different conditions for depositing tailings in Żelazny Most TSF result in a spatially variable bearing capacity of tailings. The overall prognosis of the subsoil response to the additional load required the construction of a real loading structure. The experimental embankment was formed in the southern part of the Żelazny Most TSF (Figure 2). The construction began at the beginning of 2013 and continued with intervals until June 2014.

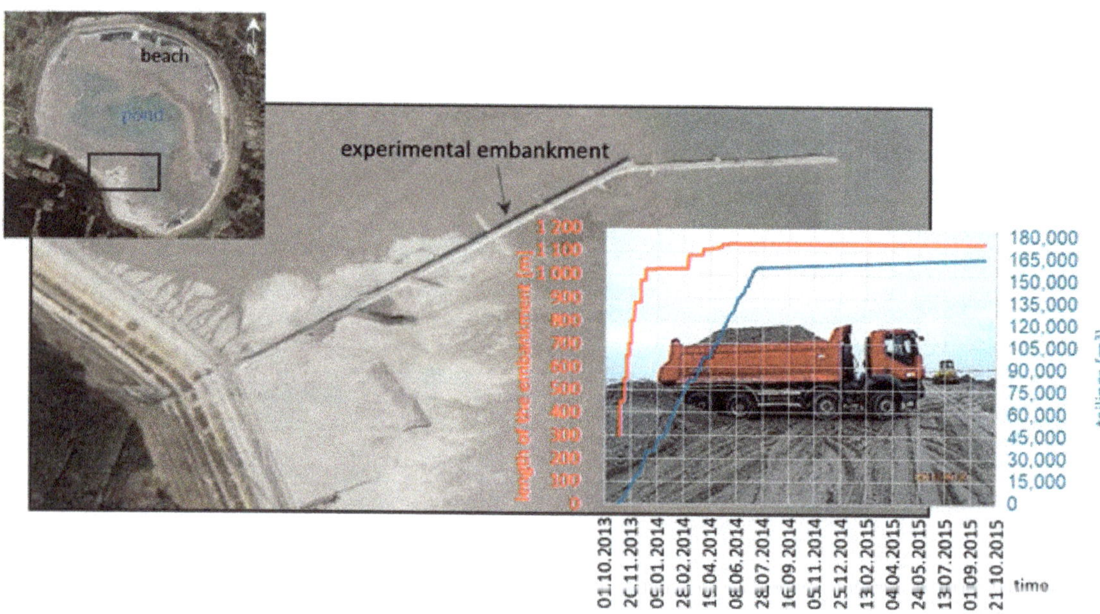

Figure 2. Experimental embankment location within the TSF and distribution of earth masses for embankment construction in time.

About 165,000 m^3 of building material were used to build the embankment. This material was selected out of postflotation tailings, generally of sandy grain size, previously deposited on the beach of the storage facility. The tailings were delivered to the construction site by road. The embankment was formed in the direction from its top part forward. The embankment, with a total length of 1140 m and a crest width of 15 m, was formed in a direction perpendicular to the axis of the facility's dam, which resulted in part of the embankment being constructed on the beach, and part within the pond, which is located in the central part of the facility (Figure 2). Analysis of Figure 2 shows that about 35,000 m^3 of waste were used to build a 1000-m-long fragment of the embankment, whereas disproportionally more waste, i.e., c. 130,000 m^3, was used to build the final 140-m-long fragment. The result of such a location was extremely variable foundation conditions and the varying bearing capacity of the embankment subsoil. Due to the location, the subsoil of the embankment was divided into 3 zones according to the scheme shown in Table 1 and in Figure 3.

Table 1. Three-zone characteristics of the subsoil of the experimental embankment.

Zone	Location	Characteristics of Embankment Subsoil
I	Beach	Zone with the highest bearing capacity and rigidity: – a direct subsoil of the embankment in the area of the dam is built with a layer of sandy tailings and a thickness of 25 m; towards the pond, the layer gets shallower until it completely disappears in the shoreline; sandy grains and the beach drainage system excludes the generation of pore water pressure excess in tailings [59], – below the profile there is a zone with a characteristic, strongly layered sandy-silty structure with varying thickness of interlayers; sandy interlayers are a filtration path that limits the formation of pore water pressure excess in silty tailings.
II	Between the beach and the pond	Transition zone: – shallowing of the sandy tailings layer towards the pond, – a visible tendency of increasing of the thickness of silty interlayers; in the area of the shoreline, the zone of strongly layered tailings reaches the highest thickness equal to the entire profile of the deposited tailings.
III	Pond	Most problematic zone: – after crossing the shoreline, the top of the silty layer is lowered creating a space that is filled by the accumulation space of tailings with grains corresponding to the finest cohesive sediments; under the bottom of the pond, the layer of the finest (silty) tailings covers the entire recognized profile, i.e., over 35 m, – a zone in a state of full saturation and with the greatest susceptibility to deformation.

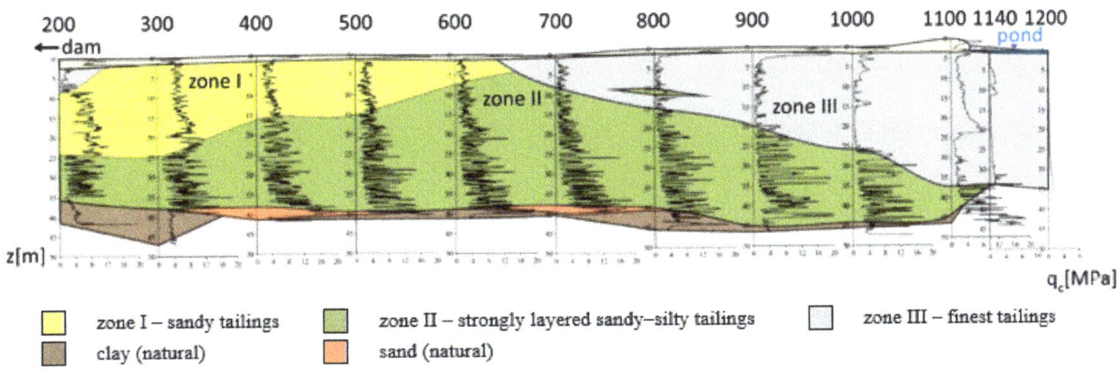

Figure 3. The embankment's subsoil (cross-section obtained on the basis of cone penetration test (CPTU) results and laboratory analysis of tailings grain size; zones as indicated in Table 1).

2.3. Data Collection

Research and observations carried out as part of the monitoring were divided into two groups (Figure 4). In situ tests included in the geotechnical monitoring (I) and geodetic monitoring (II) of real displacements of the subsoil loaded with an embankment were carried out at selected test points. Simultaneously, samples of tailings were taken for laboratory testing.

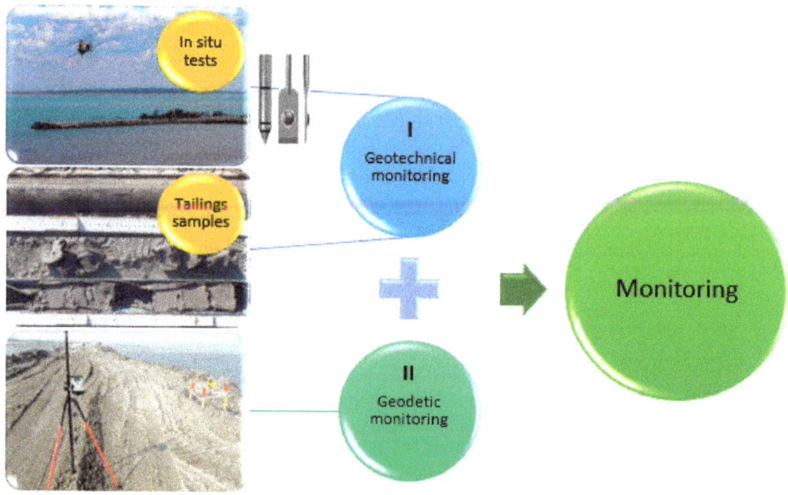

Figure 4. Monitoring components of the embankment and its subsoil.

I Geotechnical monitoring—DMT

The DMTs [60] included in the first group of tests were carried out both at the embankment crest and at a floating platform. Depending on the location of the test point, the penetration depth ranged between 12 and 42 m. DMT tests were carried out in accordance with the test standard in which two characteristic pressures, ΔA and ΔB, are determined with a frequency of 20 cm every 20-cm increase in the penetration depth. According to Marchetti [36], external pressure ΔA and internal pressure ΔB are used to correct the A and B readings (positions of the dilatometer membrane) into corrected readings, the corrected first reading p_0 and corrected second reading p_1. In the next step, intermediate parameters are determined, the material index I_D, horizontal stress index K_D, and Dilatometer modulus E_D. Then, intermediate parameters are converted by means of commonly used correlations to geotechnical parameters, e.g., constrained modulus M_{DMT}, cohesion c_u, and friction angle φ (Figure 5). M and c_u are generally the most useful and accurate parameters by DMT [61,62]. According to Marchetti [36], predicting settlements of shallow foundations is probably the number one application of the DMT. This statement applies in particular to sands where undisturbed samples cannot be retrieved. Based on M_{DMT}, settlements of the tailings loaded with an embankment were estimated. Settlements were calculated by means of the one-dimensional formula ($s_{1,DMT}$) according to the diagram shown in Figure 5. Settlements of postflotation tailings were calculated using DMT Settlements Software (v 1.0.1.16, Studio Prof. Marchetti, Rome, Italy). According to the description, the DMT Settlements Software computes the one-dimensional conventional settlements calculation below uniformly loaded surface areas of flexible loads using the DMT results. The software is designed to import from .uni files the constrained modulus of the soil and the vertical effective stress from the DMT. To perform a settlements calculation, the following input must be given: loaded area (defines the load in terms of weight and geometry), soil parameters (define the soil in terms of modulus and vertical effective stress), and calculation options (define specific parameters and criteria used in the calculation).

Vertical stress increments are evaluated according to the Boussinesq theory of elasticity for homogeneous elastic half space.

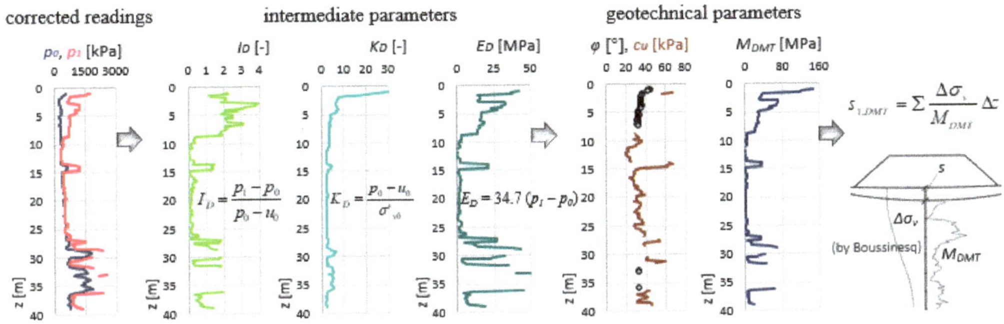

$M_{DMT} = R_M - E_D$	if:	$I_D \leq 0.6$ $I_D \geq 3.0$ $0.6 < I_D < 3.0$ where: $K_D > 10$ $R_M < 0.85$	$R_M = 0.14 + 2.36 \log K_D$ $R_M = 0.5 + 2 \log K_D$ $R_M = R_{M0} = (2.5 - R_{M0}) \log K_D$ $R_{M0} = 0.14 + 0.15 (I_D - 0{,}6)$ $R_M = 0.32 + 2.18 \log K_D$ $R_M = 0.85$

Figure 5. Scheme for settlement prediction by DMT (where: p_0 and p_1—corrected first and corrected second reading, I_D—material index, K_D—horizontal stress index, E_D—dilatometer modulus, φ—friction angle, u_0—equilibrium pore pressure, M_{DMT}—constrained modulus, σ_v—vertical stress, σ'_{v0}—effective overburden stress, z—depth, R_M—correction factor, s—settlements; data comes from the DMT performed on the embankment axis at a distance of 1 km from the TSF dam).

II Geodetic monitoring

Observed settlements were determined on the basis of data obtained from geodetic monitoring (Figure 4). In order to carry out displacement measurements, a geodetic network was established. It consists of 62 control points installed both at the top and in the immediate vicinity of the embankment. Measurements were made using Global Navigation Satellite System (GNSS), in Real Time Kinematic (RTK) and Real Time Network (RTN) measurement modes using the ASG-EUPOS system. Control measurements were carried out every two weeks throughout the entire embankment construction period and for one year after its completion. In total, 1382 geodetic measurements were carried out.

3. Results

The results of the conducted DMT and geodetic measurements enabled assessment of the settlement of tailings. Two limitations were applied while selecting test points at which the settlements were examined. First, the analysis was restricted to the results of a section of 600 m, at which test points were located every 100 m. The analysis was performed for an embankment section from the point located 400 m away from the TSF dam to the point located 1000 m from the dam. Because of minor settlement values for the stiff sandy beach of the facility, the fragment located between the dam of the TSF, and a point located 400 m away from the dam were excluded. The second limitation applied to the selection of test points which would provide credible DMT results and geodetic measurements.

3.1. Predicted Tailings Settlements—DMT

The calculated settlements are obtained using the interpretation formulae and the calculation method recommended in [45]. Cumulative settlement results are summarized in Table 2. The analysis of settlements calculation indicates that the smallest settlements are found in zone I (Table 1), where the embankment was placed on the sandy TSF beach. The settlement values in zone I are between 7.69 cm at 0+400 and 8.13 cm at point 0+500 (Table 2). At the point located 100 m away (0+600), the settlement values are already twice as large and exceed 16 cm (zone II in Tables 1 and 2). At the next point from zone II (0+700), the settlement values are reduced, reaching 5.54 cm. Lower settlement values can be explained by the fact that the embankment height at point 0+700 was lower, which resulted in lower subsoil load than in adjacent points (Table 2). At three consecutive points (zone III), settlement values increased from 56.54 cm (0+800), through 62.81 cm (0+900), reaching the highest value of 165.78 cm at point 1000 m from the TSF dam (zone III in Tables 1 and 2). At point 1+000, the largest settlement was to be expected due to the identification of a layer of very weak silty deposits, which build an unstable subsoil of the embankment at the bottom of the pond. Figure 6 shows the result sheet from the program DMT Settlements Software for the test point where the largest settlements were determined.

Table 2. Summary of results of settlements calculation, below the centre of embankment.

Zone	Location [1] (km)	Layers [2] (–)	σ_v (kPa)	s_{DMT} (cm)	Settlements Below the Centre
I	0+400	183	70	7.69	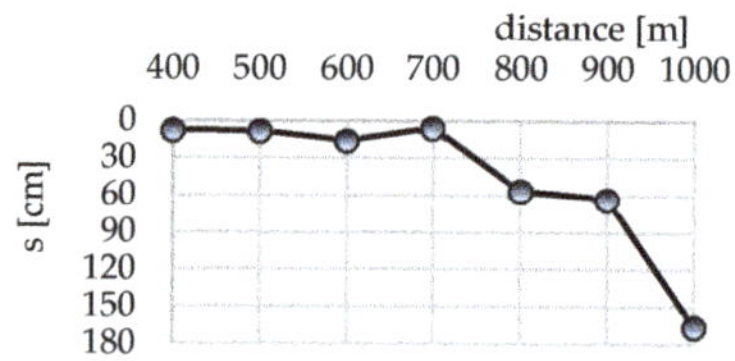
	0+500	190	75	8.13	
II	0+600	184	80	16.06	
	0+700	193	75	5.54	
III	0+800	160	80	56.54	
	0+900	169	85	62.81	
	1+000	171	93	165.78	

[1] Location (km) = km from the TSF dam. [2] Number of layers, where thickness of one calculation layer = 0.20 m.

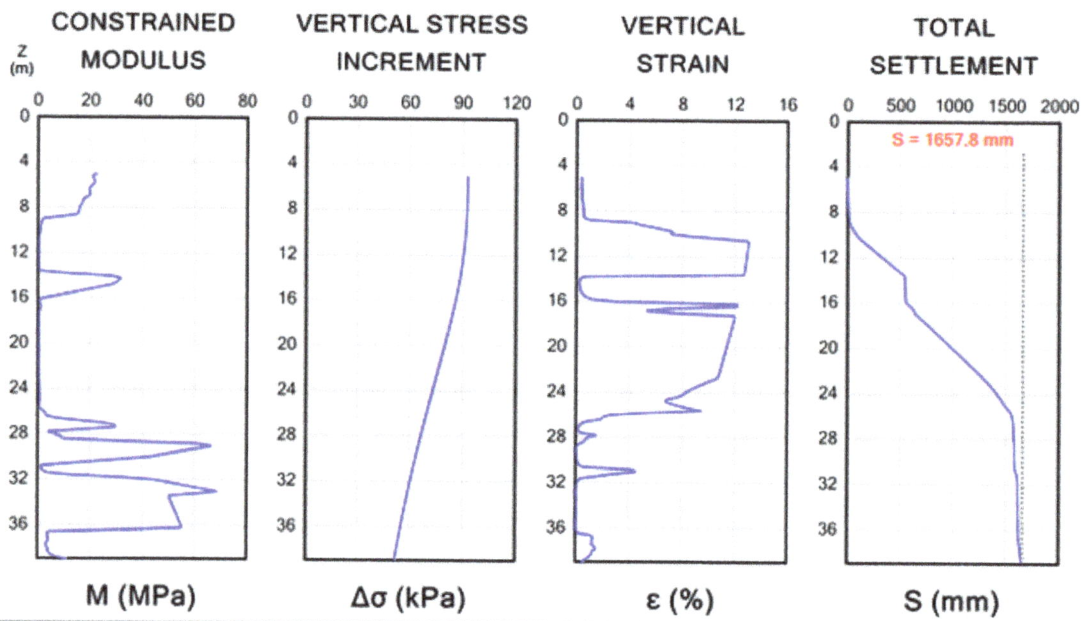

Calculation options:

– lower limit of *M* assigned in the calculation	0.70 MPa
– thickness of calculation layer	0.20 MPa
– end of calculation = end of assigned profile	39 m

Figure 6. Results sheet of settlements calculation, below the centre (test point 1+000).

3.2. Observed Tailings Settlements—Geodetic Measurements

The results of the geodetic measurements enabled the determination of the actual settlement values of the loaded subsoil. The measurements at points analogous to the DMT were performed at the control points located in the embankment's axis at its top. The same part of the 600-m-long embankment was analysed starting from a point located 400 m from the TSF dam. Cumulative settlement results are summarized in Table 3. Observed settlements in zone I (Table 3) reach values from 21.60 cm (0+500) to 30.03 cm (0+400). At the first point of zone II (0+600), the settlement value is exactly the same as in the first test point in zone I (0+400). At the second point (0+700), a decrease in the settlement value to 19.70 cm is noted. At subsequent research points (zone III), settlement values regularly increase with the distance from the TSF dam towards the pond. At a point 800 m away from the TSF dam, the settlements reach a value almost double that of the previous point, amounting to 37.51 cm, while at 0+900 they exceed 0.5 m. The highest settlement values were observed at the last analysed point: the measurement value was almost twice as high as at the previous point and equalled 92.25 cm (Table 3).

Table 3. Summary of the results of the observed settlements.

Zone	Location [1] (km)	$s_{observed}$ (cm)	Settlements Below the Centre
I	0+400	30.03	
	0+500	21.60	
II	0+600	30.03	
	0+700	19.70	
III	0+800	37.51	
	0+900	50.09	
	1+000	92.25	

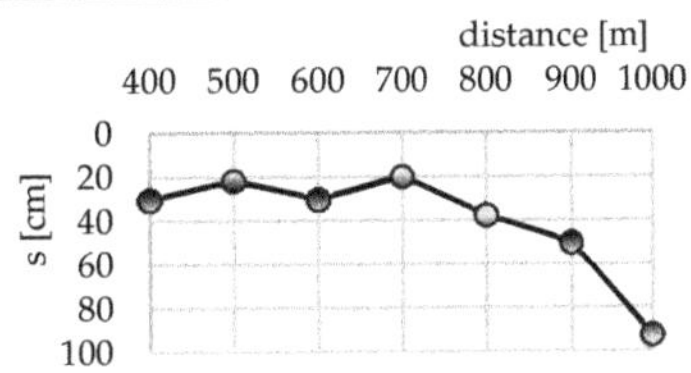

[1] Location (km) = km from the TSF dam.

3.3. The Comparison of the Predicted and Observed Settlements of Tailings

Based on the analysis of the results, it was found that settlements, both predicted by DMT and observed (geodetic measurements), in overall terms, progressively increase as they move away from the beach towards the centre of the storage facility where the pond is located. Because of the central location of the pond, which is in the zone more susceptible to deformation of the subsoil (zone III), settlements in this place reach the highest values, which are several times higher than in zone I and II (Table 4).

Table 4. Comparison of settlements.

Zone	Location [1] (km)	Settlements (cm)				Settlement Ratio
		$DMT_{predict.}$	Avg. $DMT_{predict.}$	Observed	Avg. Observed	$s_{DMT}/s_{observed}$
I	0+400	7.69	7.91	30.03	25.82	0.26
	0+500	8.13		21.60		0.38
II	0+600	16.06	10.80	30.03	24.87	0.53
	0+700	5.54		19.70		0.28
III	0+800	56.54	95.04	37.51	59.95	1.51
	0+900	62.81		50.09		1.25
	1+000	165.78		92.25		1.80

[1] Location (km) = km from the TSF dam.

When analysing settlements divided into individual zones, it should be stated that predicted and observed settlements differ in each of these zones. With definitely different means of the predicted and observed settlements in zone I, respectively 7.91 cm and 25.82 cm, settlement ratios ($s_{DMT}/s_{observed}$) equalled 0.26 and 0.38 (Table 4). A slightly greater settlement compliance was demonstrated in zone II, as the average settlements predicted by DMT are 10.80 cm, while the average observed settlements is 24.87 cm, while at 0+600 predicted settlements were twice smaller than the observed settlements, $s_{DMT}/s_{observed}$ = 0.53 (Table 4). In the next point in zone II (0+700), predicted settlements were 5.54 cm with four times the observed settlements, $s_{DMT}/s_{observed}$ = 0.28. The lower settlement value at test point 0+700, which was explained in 3.1, locally contradicts the overall trend of an increase in the settlement value towards the pond. Although the settlements determined using both methods differ significantly, it can be seen that the lower settlement values at a point 700 m away from the TSF dam were obtained both on the basis of the results of DMT and geodetic measurements. This fact confirms that a lower load (smaller embankment height) had a direct impact on the lower settlement value at point 0+700. At the next three test points (zone III), both predicted and observed settlements increase in line with the overall trend (Table 4, Figure 7a). At point 0+800, the settlement ratio is 1.5, which is also the average value of this ratio for the entire zone III. The settlement ratio at the 0+900 test point equals 1.25. This value indicates the highest convergence in the value of settlements obtained by the two methods considered. At the last test point (1+000) $s_{DMT}/s_{observed}$ is already 1.80. The main impact on the average settlements in zone III of 95.04 cm (predicted) and 59.95 cm (observed) was due to the large discrepancy of results at the last analysed point 1+000, where predicted settlements amounted to 165.78 cm, while observed equalled 92.25 cm. At this point, tailings are characterized by the largest anisotropy resulting from the cyclic spigotting of postflotation sediments migrating to the pond, where the accumulation of postflotation sediments of the finest grain structure is increased. In addition, it should be noted that 30 m further from the last analysed test point, the observed settlements were already 156.63 cm, i.e., comparable to the predicted settlements (165.78 cm) at 1+000 ($s_{DMT(1+030)}/s_{observed(1+1000)}$ = 1.06). However, there is no direct possibility to compare the results due to the fact that the 1+030 point is an intermediate point where the DMT test was not performed. A comparison of the predicted and observed settlements and settlements agreement is shown in Figure 7.

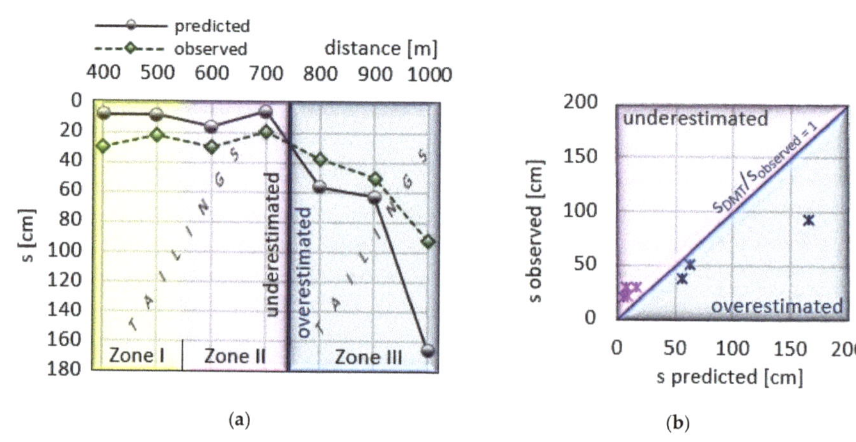

Figure 7. (**a**) Predicted and observed settlements comparison; (**b**) settlements agreement of tailings.

4. Discussion

Unstructured postflotation tailings, after entering the facility, are subject to the deposition process. By analogy to the natural deposition process that occurs as a result of energy being lost by the stream carrying the hydrated material, postflotation sediments begin to freely sediment inside the storage facility. As a result of tailings particle sedimentation, postflotation sediments are formed. Close to the outlets of the spigotting pipes (beach = zone I) are the thickest particles, most often of sandy grain size, which potentially exhibit the properties of fine-grained non-cohesive soil. The further from the place of waste dumping, the more layered the sediment is, and individual layers are characterized by a smaller thickness. The volume of waste of finer grain size dominated by the silty fraction also increases. Particles with the finest grain size (silt and clay) migrate to the pond. Postflotation sediments, which freely sediment in the salty water environment, combine into porous aggregates with a strongly layered structure. The effects of natural sedimentation segregation are changes in the physical and mechanical properties of the waste as a function of the distance from the place of discharge. These changes directly affect the response of tailings to additional load in individual zones (Table 1).

The most reliable way to assess the actual response seems to be geodetic measurements of settlements. However, these are long-term measurements on the basis of which total settlements cannot be estimated as a result of a single measurement. Geotechnical in situ tests, such as DMTs, are successfully used to predict the settlement of natural soils [36,40,43,49]. In the case under consideration, the values of tailings settlements predicted by DMTs differ from the observed settlements. Some results are overestimated, while some are underestimated. A clear underestimation of settlements takes place in zone I, where the subsoil is built of sandy layers (beach) characterized by a higher value of constrained moduli. Settlements calculated based on overestimated constrained moduli should be underestimated [63]. A slightly better fit is observed in zone II, separating the sandy beach from the pond. In zone III, on the other hand, where the embankment subsoil, which is also the bottom of the pond, is built of low-bearing postflotation sediments, predicted settlements are overestimated. A similar problem is very common. Settlement values of natural soils for the Barcelona airport terminal determined based on CPTU and DMT results showed an overestimation of settlements by 28% compared to measured settlements [29]. However, the authors argued that the method used supplemented by a local correlation (M_{CPTU} and M_{DMT}) can be successfully used to estimate settlements. According to Schmertmann [32], the average ratio of natural soils settlement predicted by DMT to actual observations was 1.18, ranging from 0.7 to 1.3. [38], also obtained high agreement between settlements predicted by DMT and measured settlements. The average ratio of settlement results determined from DMT tests to actual observations equalled 1.87, ranging from about 1.0 to 2.5. In the case of an embankment built within the Żelazny Most TSF, the settlement ratio is in the range of 0.26 to 1.8, with an average value of 0.86. However, using the average settlement ratio is not justified in this case. Consequently, this leads to far-reaching average values. The use of average settlement values predicted by DMT and observed (46.08 cm and 40.17 cm, respectively) looks similar. It should be remembered that the loaded subsoil is built of postflotation tailings, not natural soil. Characteristics of the latter are much better recognized, and in addition, they do not have such specific zoning, unlike tailings. Therefore, it will be more appropriate, if at all, to use average values within separate zones (Table 1), where the average settlement ratio is 0.31 in zone I, 0.43 in zone II, and 1.59 in zone III. This approach clearly indicates underestimation of settlements in the first two zones and overestimation in the last zone (Figure 7). The discrepancy of results can be explained in two ways: first of all, paying attention to the DMT test interval, which may result in an increased frequency of random soil parameters registration, with higher stiffness (heavily layered sand and silt tailings of zones I and II, underestimated settlements) and weak zone III waste (overestimated settlements). Despite this, the literature emphasizes that the measurement interval along with the speed of the DMT test results in the fact that, in addition to determining soil parameters, predicting settlements is the number one application of the DMT, as mentioned earlier [41,47]. The results of other geotechnical tests at a larger interval, performed to assess settlements (e.g., SPT, standard penetration test, generally performed at 1.5-m intervals), are considered to be unreliable and not credible [37,42]. Settlements predicted by SPT can be overestimated against settlements predicted by DMT [64,65]. Other tests, such as PMT (pressuremeter test) and SBPM (self-boring pressuremeter), are time-consuming and less economical [43,46]; secondly, the results of the DMT do not include immediate and secondary settlements, which could partly be included in the results of geodetic measurements (zone I and II). It should be noted that the predicted settlements are meant to be "the settlement in working conditions". According to Monaco et al. [46], the DMT-calculated settlement is the primary settlement. A similar situation occurred in the case of the experimental embankment at Treporti (Venice, Italy) [41,46,47,66]. According to Marchetti et al. [41], measured settlements amounted to 360 mm (including secondary), while settlements predicted by DMT reached 290 mm (net of secondary). The discrepancy of the results was explained by not including part of the secondary settlements (70 mm, which is less than 20%). The adulators of [30,35,38,42,44,49,67] also indicate high agreement of values predicted by DMT and observed settlements. In Żelazny Most TSF, more similar results of settlement tailings predicted by DMT in relation to the results determined on the basis of geodetic measurements may be obtained in the future by introducing local correlations to determine the constrained modulus M (e.g., correlations based on the results of DMT and CPTU).

5. Conclusions

Considering the results of the conducted research and the results available in the literature, it can be stated that the DMT test, widely recommended for determining settlement of natural soils, can also be used to estimate the settlement trend of tailings. The results of the DMT tests allow the determination of approximate tailings settlement values. In addition, they allow determination of the overall trend of settlement of weak tailings in line with the trend determined on the basis of geodetic observations. The conducted analyses led to the following general conclusions being drawn:

- Settlement of the postflotation tailings, depending on the facility zone, range from a few cm within the beach of the facility to over 1.5 m within the supernatant pond;
- Based on linear elasticity, DMT tests provide a settlement proportional to the load (Table 2); the results of the DMT test obtained after applying the load allow an approximate determination of the actual response of the tailings to the additional load; they indicate the trend of settlement consistent with the trend determined on the basis of geodetic measurements (Figure 7a);
- The results of DMT and geodetic measurements document the general correctness of the increase in the value of tailings settlement as they move away from the facility's dam towards the pond; predicted and observed settlements in zone I and II < predicted and observed settlements in zone III (Table 4, Figure 7a);
- The DMT in relation to geodetic measurements clearly underestimates (zone I and II in Table 1) settlements of a more rigid subsoil and vice versa; overestimates (zone III in Table 1) settlements after crossing the pond line (Table 4, Figure 7b); and
- The weakness of DMT in the context of strongly layered tailings may be the aspect of discontinuous measurement, which in turn may lead to random measurements; the choice of a dilatometer membrane with appropriate stiffness, tailored to the tailings characteristics, may also affect the recorded test parameters. Omission of this element may directly affect the credibility of results obtained from the DMT test. The weakness of geodetic measurements is the time in which measurements are made and their cyclicality, which is next to the unification of methodology and interpretation of results, one of the three principles of monitoring.

Funding: The publication was co-financed within the framework of Ministry of Science and Higher Education programme as "Regional Initiative Excellence" in years 2019–2022, Project No. 005/RID/2018/19.

Data Availability Statement: Data sharing is not applicable to this article.

Conflicts of Interest: The authors declare no conflict of interest.

References

1. Jamiolkowski, M.; Carrier, W.D.; Chandler, R.J.; Høeg, K.; Świerczyński, W.; Wolski, W. The geotechnical problems of the second world largest copper tailings pond at Zelazny Most, Poland. In *1st Za Chieh-Moh Distinguished Lecture keynote speech, Proceedings of the 17th SEAGC South East Asian Geotechnical Conference, Taipei, Taiwan, 10–13 May 2010*; Li, J.C.C., Lin, M.L., Eds.; Taiwan Geotechnical Society: Taipei, Taiwan, 2010; Volume 2, pp. 12–27.
2. Jamiolkowski, M. Soil mechanics and the observational method: Challenges at the Zelazny Most copper tailings disposal facility. *Geotechnique* **2014**, *64*, 590–618. [CrossRef]
3. Świdziński, W.; Tschuschke, W.; Świerczyński, W.; Wolski, W. Tailings Storage Facility Żelazny Most—Significant geotechnical challenge. *Mar. Eng. Geotech.* **2015**, *36*, 186–193.
4. Stefaniak, K.; Wróżyńska, M. On possibilities of using global monitoring in effective prevention of tailings storage facilities failures. *Environ. Sci. Pollut. Res.* **2018**, *25*, 5280–5297. [CrossRef] [PubMed]
5. Lambrechts, J.R.; Leonards, G.A. Effects of stress history on deformation of sand. *J. Geotech. Geoenviron.* **1978**, *104*, 1371–1387.
6. Abbiss, C.P. Calculation of elasticities and settlements for long periods of time and high strains from seismic measurements. *Geotechnique* **2014**, *33*, 397–405. [CrossRef]
7. Campanella, R.G.; Robertson, P.K. A seismic cone penetrometer to measure engineering properties of soil. In Proceedings of the 54th Annual International Meeting and Exposition of the Society of Exploration Geophysics, Atlanta, GA, USA, 6–7 December 1984; pp. 138–141.
8. Kaggwa, W.S.; Jha, R.K.; Jaksa, M.B. Use of Dilatometer and Cone Penetration Tests to Estimate Settlements of Footings on Calcareous Sand. In Proceedings of the 7th Australia New Zealand Conference on Geomechanics: Geomechanics in a Changing World, Adelaide, Australia, 1–5 July 1996; pp. 909–914.
9. Tezcan, S.S.; Keceli, A.; Ozdemir, Z. Allowable bearing capacity of shallow foundations based on shear wave velocity. *Geotech Geol. Eng.* **2006**, *24*, 203–218. [CrossRef]
10. Mayne, P.W. Georgia Institute of Technology. In *Cone Penetration Testing State-of-Practice*; NCHRP Project; Transportation Research Board, National Academies Press: Washington, DC, USA, 2007.
11. Robertson, P.K. Interpretation of cone penetration tests—A unified approach. *Can. Geotech. J.* **2009**, *46*, 1337–1355. [CrossRef]
12. Liu, S.; Cai, G.; Puppala, A.J.; Tu, Q. Prediction of embankment settlements over marine clay using piezocone penetration tests. *Bull. Eng. Geol. Environ.* **2011**, *70*, 401–409. [CrossRef]
13. Omar, M.N.; Abbiss, C.P.; Taha, M.R.; Nayan, K.A.M. Prediction of long-term settlement on soft clay using shear wave velocity and damping characteristics. *Eng. Geol.* **2011**, *123*, 259–270. [CrossRef]
14. L'Heureux, J.S.; Long, M.; Vanneste, M.; Sauvin, G.; Hansen, L.; Polom, U.; Lecomte, I.; Dehls, J.; Janbu, N. On the prediction of settlement from high-resolution shear-wave reflection seismic data: The Trondheim harbour case study, mid Norway. *Eng. Geol.* **2013**, *167*, 72–83. [CrossRef]
15. Tonni, L.; Gottard, G. Assessing compressibility characteristics of silty soils from CPTU: Lessons learnt from the Treporti Test Site, Venetian Lagoon (Italy). *AIMS Geosci.* **2019**, *5*, 117–144. [CrossRef]
16. Li, Y.; Shen, Y.; Wang, X.; Li, S.; Li, T.; Zhao, Q. Theoretical Studies and Implementation on the Temporary Data Storage Method for Cone Penetration Test. *Sensors* **2021**, *21*, 575. [CrossRef]
17. Mayne, P.W. In-situ test calibrations for evaluating soil parameters. In *Characterization and Engineering Properties of Natural Soils*; Tan, T.S., Phoon, K.K., Hight, D.W., Leroueil, S., Eds.; Taylor & Francis Group: London, UK, 2007; Volume 3, pp. 1601–1652.
18. Abu-Farsakh, M.; Yu, X. Comparison of predicted embankment settlement from piezocone penetration test with field measurement and laboratory estimated. In *Geotechnical and Geophysical Site Characterization 4*; Coutinho, R.Q., Mayne, P.W., Eds.; Taylor & Francis Group: London, UK, 2013; pp. 349–355.
19. Drevininkas, A.; Creer, G.; Nkemitag, M. Comparison of estimated settlements from CPTu and SDMT in organic soils. In *Geotechnical and Geophysical Site Characterization 4*; Coutinho, R.Q., Mayne, P.W., Eds.; Taylor & Francis Group: London, UK, 2013; pp. 1839–1846.
20. Bastani, S.A.; Silver, G.P.; Atkinson, D.R. CPT based settlement prediction over California soft rock. In Proceedings of the 3rd International Symposium on Cone Penetrometer Testing, CPT'14, Las Vegas, NV, USA, 12–14 May 2014; pp. 1113–1120.
21. Long, M.; Carrol, R. Application of CPTU and SDMT to the characterization of Irish silts. In *5th International Workshop CPTU and DMT in Soft Clays and Organic Soils*; Młynarek, Z., Wierzbicki, J., Eds.; Polish Committee on Geotechnics: Poznan, Poland, 2014; pp. 45–63.
22. Shkodrani, N.; Anamali, E.; Dhimitri, L.; Gashi, S. In Situ Tests—Predicted vs. Observed Settlements: A Comparative Case Study. *J. Civil Eng. Archit.* **2014**, *8*, 824–832. [CrossRef]
23. Tschuschke, W.; Kumor, M.K.; Walczak, M.; Tschuschke, M. Cone penetration test in assessment of soil stiffness. *Geol. Q.* **2015**, *59*, 419–425. [CrossRef]
24. Robertson, P.K.; Campanella, R.G.; Gillespie, D.; Greig, J. Use of Piezometer Cone data. In *Use of In-Situ Testing in Geotechnical Engineering*; In-Situ'86, GSP 6; ASCE: Reston, VA, USA, 1986; pp. 1263–1280.
25. Jamiolkowski, M.; Ghionna, V.; Lancellotta, R.; Pasqualini, E. New corrections of penetration tests for design practice. In Proceedings of the 1st International Symposium on Penetration Testing, ISOPT I, Orlando, FL, USA, 20–24 March 1988; Balkema: Orlando, FL, USA, 1988; Volume 1, pp. 263–296.

26. Robertson, P.K.; Cabal Robertson, K.L. *Guide to Cone Penetration Testing for Geotechnical Engineering*, 6th ed.; Gregg Drilling & Testing: Signal Hill, CA, USA, 2015.
27. Been, K.; Quiñonez, A.; Sancio, R.B. Interpretation of the CPT in engineering practices. Key note lecture 2. In Proceedings of the 2nd International Symposium on Cone Penetration Testing, Huntington Beach, CA, USA, 9–11 May 2010; Robertson, P.K., Ed.; Omnipress: Huntington Beach, CA, USA, 2010.
28. Lunne, T.; Robertson, P.K.; Powell, J. *Cone Penetration Testing in Geotechnical Practice*; Blackie Academic and Professional, CRC Press: London, UK, 1997.
29. Arroyo, M.M.; Mateos, M.T.; Devincenzi, M.; Gómez-Escoubès, R.; Martínez, J.M. CPTu-DMT performance-based correlation for settlement design. In *2nd International Conference on Geotechnical Site Characterization ISC'2*; Viana da Fonseca, A., Ed.; Millpress: Porto, Portugal, 2004; pp. 1605–1611.
30. Mayne, P.W. Lateral drilled shaft response from dilatometer tests. In *GeoSupport Conference 2004: Drilled Shafts, Micropiling, Deep Mixing, Remedial Methods and Specialty Foundation Systems*; Geotechnical Special Publication; Turner, J.P., Mayne, P.W., Eds.; American Society of Civil Engineers: Orlando, FL, USA, 2004. [CrossRef]
31. Lacasse, S.; Lunne, T. Dilatometer Tests in Sand. In *In Situ '86, ASCE Spec. Conf. on Use of In Situ Tests in Geotechnical Engineering*; ASCE Geotech. Spec. Publ.; Virginia Tech.: Blacksburg, VA, USA, 1986; Volume 6, pp. 686–699.
32. Schmertmann, J.H. Dilatometer Tests in Sand. In *In Situ '86, ASCE Spec. Conf. on Use of In Situ Tests in Geotechnical Engineering*; ASCE Geotech. Spec. Publ.; Virginia Tech.: Blacksburg, VA, USA, 1986; Volume 6, pp. 303–321.
33. Leonards, G.A.; Frost, J.D. Settlements of Shallow Foundations on Granular Soils. *J. Geotech. Eng.* **1988**, *114*, 791–809. [CrossRef]
34. Marchetti, S. Discussion to Leonards, G.A. & Frost, J.D. Settlements of Shallow Foundations on Granular Soils (ASCE Jnl GE, 1988, 114). *ASCE J. Geotech. Eng.* **1991**, *117*, 174–179.
35. Marchetti, S. The Flat Dilatometer: Design Applications. Keynote Lecture. In Proceedings of the 3rd Third International Geotechnical Engineering Conference, Cairo, Egypt, 5–8 January 1997; pp. 421–448.
36. Marchetti, S. Flat Dilatometer (DMT). Applications and Recent Developments. In Proceedings of the 50th Indian Geotechnical Conference IGC-2015, Pune, India, 16–19 December 2015; pp. 83–92.
37. Woodward, M.B.; McIntosh, K.A. Case history: Shallow Foundation Settlement Prediction Using the Marchetti Dilatometer. In Proceedings of the ASCE Annual Florida Section Meeting, Orlando, FL, USA, 7–9 June 1993.
38. Skiles, D.L.; Townsend, F.C. Predicting Shallow Foundation Settlement in Sands from DMT. In *Vertical and Horizontal Deformations of Foundations and Embankments*; ASCE: Austin, TX, USA, 1994; Volume 40, pp. 132–142.
39. Steiner, W. Settlement Behavior of an Avalanche Protection Gallery Founded on Loose Sandy Silt. In *Vertical and Horizontal Deformations of Foundations and Embankments*; ASCE: Austin, TX, USA, 1994; Volume 40, pp. 207–221.
40. Sawada, S.; Sugawara, N. Evaluation of densification of loose sand by SBP and DMT. In Proceedings of the 4th International Symposium: Pressuremeter and its New Avenues, Sherbrooke, QC, Canada, 17–19 May 1995; pp. 101–107.
41. Marchetti, S.; Monaco, P.; Calabrese, M.; Totani, G. DMT-predicted vs measured settlements under a full-scale instrumented embankment at Treporti (Venice, Italy). In Proceedings of the 2nd International Conference on Site Characterization ISC'2, Porto, Portugal, 19 September 2004; Volume 2, pp. 1511–1518.
42. Mayne, P.W. Unexpected but foreseeable mat settlements of Piedmont residuum. *J. Geoeng. Case Hist.* **2005**, *1*, 5–17.
43. Schnaid, F.; Ortigao, J.A.R.; Mántaras, F.M.; Cunha, R.P.; MacGregor, I. Analysis of self-boring pressuremeter (SBPM) and Marchetti dilatometer (DMT) tests in granite saprolites. *Can. Geotech. J.* **2000**, *37*, 796–810. [CrossRef]
44. Tice, J.A.; Knott, R.A. Geotechnical Planning, Design and Construction for the Cape Hatteras Light Station Relocation. *Geo-Strat. Geo-Inst. ASCE* **2000**, *3*, 18–23.
45. Marchetti, S.; Monaco, P.; Totani, G.; Calabrese, M. *The Flat Dilatometer Test (DMT) in Soil Investigations. A Report by the ISSMGE Committee TC16*. Ground Property Characterisation from In–situ Testing; *Proc. In Situ*; 2001, p. 41. Reprinted in *Proc. DMT*; 2006, Washington, DC, USA. Available online: https://www.marchetti-dmt.it/conference/dmt15/papers%20DMT%202015%20(pdf)/TC16%20DMT%20Report%202001.pdf (accessed on 29 March 2021).
46. Monaco, P.; Totani, G.; Calabrese, M. DMT—Predicted vs observed settlements: A review of the available experience. In Proceedings of the 2nd International Conference on the Flat Dilatometer, Flat Dilatometer Testing, Washington, DC, USA, 2–5 April 2006; pp. 244–252.
47. Amoroso, S. DMT—Predicted vs. observed settlements: Treporti test embankment. In Proceedings of the 2nd International Balkans Conference on Challenges of Civil Engineering, BCCCE, Tirana, Albania, 17 May 2013; pp. 793–801.
48. Buselli, F. Use of DMT to predict settlement of shallow foundations. In *Geotechnics for Sustainable Development*; Phung, D.L., Ed.; Construction Publishing House: Hanoi, Vietnam, 2013; pp. 531–536.
49. Failmezger, R.; Till, P.; Frizzell, J.; Kight, S. Redesign of shallow foundations using dilatometer tests-more case studies after DMT'06 conference. In Proceedings of the 3rd International Conference on the Flat Dilatometer, Rome, Italy, 15–17 June 2015; pp. 83–91.
50. Santos, R.; Rocha, B.; Giacheti, H. DMT for Load-Settlement Curve Prediction in a Tropical Sandy Soil Compared to Plate Load Tests. *Geotech. Test. J.* **2020**, *43*, 113–131. [CrossRef]
51. Hayes, J.A. The Marchetti Dilatometer and Compressibility. In *In Situ Testing and Monitoring*; Southern Ontario Section of Canad. Geotech. Society: Toronto, ON, Canada, 1990; pp. 1–23.

52. Massarsch, K.R. Settlement analysis of compacted granular fil. In *International Conference on Soil Mechanics and Foundation Engineering—XIII ICSMFE*; Balkema: New Delhi, India, 1994; Volume 1, pp. 325–328.
53. Tschuschke, W.; Gogolik, S.; Wróżyńska, M.; Kroll, M.; Stefanek, P. The Application of the Seismic Cone Penetration Test (SCPTU) in Tailings Water Conditions Monitoring. *Water* **2020**, *12*, 737. [CrossRef]
54. Tschuschke, W.; Wróżyńska, M.; Wierzbicki, J. Quality control for the construction of a tailings dam. *Acta Geotech. Slov.* **2017**, *14*, 3–9.
55. Stefaniak, K.; Wróżyńska, M.; Kroll, M. Application of postflotation tailings in hydroengineering structures. *J. Ecol. Eng.* **2017**, *18*, 113–118. [CrossRef]
56. Li, S.; Guo, Z.; Pan, J.; Zhu, D.; Dong, T.; Lu, S. Stepwise Utilization Process to Recover Valuable Components from Copper Slag. *Minerals* **2021**, *11*, 211. [CrossRef]
57. Robinsky, E.I. Thickened discharge—A new approach to tailings disposal. *CIM Bull.* **1975**, *68*, 47–53.
58. Robinsky, E.I. Tailings disposal by the thickened discharge method for improve economy and environmental control. *Tailing Dispos. Today* **1979**, *2*, 75–91.
59. Kroll, M.; Stefaniak, K.; Walczak, M. Determination of efficiency of the circumferential drainage system. *J. Ecol. Eng.* **2015**, *45*, 68–74. [CrossRef]
60. *ISO TS 22476-11 Geotechnical Investigation and Testing—Field Testing—Part 11: Flat Dilatometer Test*. 2017. Available online: https://www.iso.org/standard/66434.html (accessed on 29 March 2021).
61. Marchetti, S. Some 2015 Updates to the TC16 DMT Report 2001. In Proceedings of the 3rd International Conference on the Flat Dilatometer, Rome, Italy, 14–17 June 2015; pp. 43–68.
62. Marchetti, S.; Monaco, P. Short course on Flat Dilatometer (DMT). ISSMGE Committee TC16: DMT in soil investigation. In Proceedings of the Bali, Indonesia Insitu Conference, Bali, Indonesia, 21–24 May 2001; pp. 1–77.
63. Maček, M.; Smolar, J.; Petkovšek, A. The reliability of CPTu and DMT for the mechanical characterization of soft tailings. *Bull. Eng. Geol. Environ.* **2019**, *78*, 2237–2252. [CrossRef]
64. Failmezger, R.A.; Rom, D.; Ziegler, S.B. Behavioral characteristics of residual soils. SPT?—A better approach to site characterization of residual soils using other in-situ tests. In *ASCE Geotech. Special Publication*; Edelen, B., Ed.; ASCE: Reston, VA, USA, 1999; Volume 92, pp. 158–175.
65. Failmezger, R.A. Discussion to Duncan, J.M., Factor of safety and reliability in geotechnical engineering. *ASCE J. Geotech. Geoenviron. Eng.* **2001**, *127*, 703–704. [CrossRef]
66. McGillivray, A.; Mayne, P.W. Seismic piezocone and seismic flat dilatometer tests at Treporti. In *Proc. ISC-2 on Geotechnical and Geophysical Site Characterization*; Viana da Fonseca, A., Mayne, P.W., Eds.; Millpress: Rotterdam, The Netherlands, 2004; Volume 2, pp. 1695–1700.
67. Failmezger, R.A.; Bullock, P.J. Owner Involvement—Choosing Risk Factors for Shallow Foundations. In *Georisk 2011*; ASCE: Atlanta, GA, USA, 2012. [CrossRef]

minerals

MDPI

Article

Analysis of Strength and Microstructural Characteristics of Mine Backfills Containing Fly Ash and Desulfurized Gypsum

Zhiqiang Huang [1,2], Erol Yilmaz [3] and Shuai Cao [1,2,*]

1 State Key Laboratory of High-Efficient Mining and Safety of Metal Mines of Ministry of Education, University of Science and Technology Beijing, Beijing 100083, China; hzhiqiang1997@126.com
2 School of Civil and Resources Engineering, University of Science and Technology Beijing, Beijing 100083, China
3 Geotechnical Division, Department of Civil Engineering, Recep Tayyip Erdogan University, Fener, Rize TR53100, Turkey; erol.yilmaz@erdogan.edu.tr
* Correspondence: sandy_cao@ustb.edu.cn

Abstract: The utilization of solid wastes (SWs) as a potential resource for backfilling is not only conducive to environmental protection but also reduces the surface storage of waste. Two types of SWs, including fly ash (FA) and desulfurized gypsum (DG), were used to prepare cementitious backfilling materials for underground mined-out areas. Ordinary Portland cement (OPC) was used as cement in mine backfill. To better investigate the feasibility of preparing backfill materials, some laboratory tests, such as uniaxial compressive strength (UCS), scanning electron microscopy (SEM), and energy dissipation theory, were conducted to explore both strength and microstructural properties of backfilling. Results have demonstrated that the main components of FA and DG in this study are oxides, with few toxic and heavy metal components. The ideal ratio of OPC:FA:DG is 1:6:2 and the corresponding UCS values are 2.5 and 4.2 MPa when the curing time are 7 days and 14 days, respectively. Moreover, the average UCS value of backfilling samples gradually decreased when the proportion of DG in the mixture increased. The main failure modes of various backfilling materials are tensile and shearing cracks. In addition, the corresponding relations among total input energy, dissipated energy and strain energy, and stress–strain curve were investigated. The spatial distribution of oxygen, aluminum, silicon, calcium, iron and magnesium elements, and hydration product are explored from the microstructure's perspective. The findings of this study provide both invaluable information and industrial applications for the efficient management of solid waste, based on sustainable development and circular economy.

Keywords: solid waste; sustainable waste management; strength characteristics; microstructural analysis; backfilling materials

Citation: Huang, Z.; Yilmaz, E.; Cao, S. Analysis of Strength and Microstructural Characteristics of Mine Backfills Containing Fly Ash and Desulfurized Gypsum. *Minerals* **2021**, *11*, 409. https://doi.org/10.3390/min11040409

Academic Editors: Abbas Taheri, Chongchong Qi, Guichen Li, Hakan Basarir, Qiusong Chen and Yuantian Sun

Received: 14 March 2021
Accepted: 8 April 2021
Published: 13 April 2021

1. Introduction

The extraction of ore resources has promoted the progress of human civilization, providing critical raw materials for sustainable development and circular economy [1,2]. However, it is irrefutable that resource mining has also caused a series of complications, such as water and air pollution [3], geotechnical risks [4], environmental hazards [5], farmland occupation [6], and tailings dam failures [7]. Solid waste output is growing at an annual rate of over 1 billion tons [8]. With the environmental protection policies and sustainable development goals, the whole use of solid waste has become a key strategy to building a resource-saving and environment-friendly city worldwide [9]. Solid wastes may cause some geotechnical and chemical instabilities if they are not well managed.

Wastes like fly ash (FA) and desulfurization gypsum (DG) are two main industrial solid wastes in Baotou city in Inner Mongolia, China [10]. Improving the utilization rate of FA and DG is one of the important technical problems to be solved by the industry [11]. At the moment, FA is mainly used as an additive for cement and the preparation of cementitious materials, while DG is mainly used to prepare gypsum-based new building materials [12,13]. Numerous scholars and engineers have conducted the feasibility of the above two types of solid waste utilization [14,15]. It was found that the risk assessment codes in FA posed no or low risk [16], while the heavy metals in DG showed a medium risk to the ecosystem [17]. In addition, the lime, FA, and DG were chosen as additives for investigating the solidification and stabilization of swelling soils [18]. Wang et al. [19] found that proper dosage and mineral additions affect the mechanical strength of the flue gas DG and FA system. Moreover, it was found experimentally that gypsum-based composites' properties improved due to the synergistic effect between industrial wastes and straw fibers [20].

The existence of underground mined-out areas seriously threatened workers and equipment; on the other hand, it severely restricts the reform of mining methods [21]. Especially for mines where there are toxic elements (e.g., cyanide, arsenic, and lead) in the ore itself, this effect is even more obvious [22]. Tailings are called solid waste or hazardous waste after ore processing [23,24]. They can also be prepared concrete additives, backfilling, and construction materials [25,26]. However, tailings containing toxic elements or heavy metal ions cannot be used for filling underground voids according to the related environmental protection standards [27–29]. Thus, exploring the feasibility of using FA and DG to prepare mine backfilling is the main purpose of this study. Scholars have so far used backfill samples with ecofriendly binders [30,31] and cement-based composites [32] on a single or combination of FA [33,34] and DG [35,36] as industrial waste.

In this study, cement, FA, and DG are utilized to prepare the backfill materials for underground voids. A series of laboratory tests, including uniaxial compressive strength, energy dissipation theory, and scanning electron microscopy, are conducted. The present study investigates the particle size distribution and chemical composition of FA and DG, strength and loading energy development, and cement-based backfills' microstructural characteristics. This study's main contribution is that it can provide a reliable reference for the feasibility of preparing mine backfills prepared with solid waste.

2. Materials and Methods

2.1. Materials

2.1.1. Characteristics of FA and DG

In this study, two solid wastes (e.g., FA and DG) obtained from Inner Mongolia were selected as additive materials for preparing the tested mine backfill specimens. Particle size distributions of these two materials were measured by the LSPOP (9) laser particle size analyzer (OMEC, Zhuhai, China). Figure 1 shows the particle size distributions of both FA and DG materials. Moreover, the physical properties of FA and DG materials were shown in Table 1.

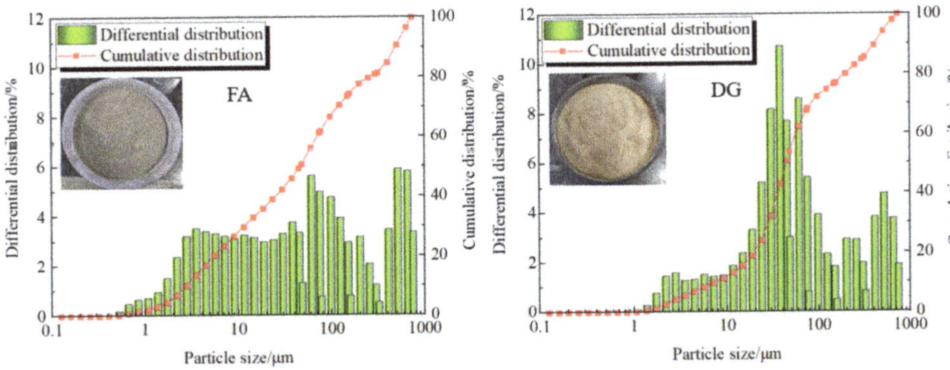

Figure 1. Particle size distributions of fly ash (FA) and desulfurization gypsum (DG) materials.

Table 1. Physical parameters of FA and DG materials.

Materials	D10 (μm)	D25 (μm)	D50 (μm)	D75 (μm)	D90 (μm)	Specific Surface Area (m^2/kg)
FA	2.766	8.067	46.324	163.687	499.833	737.135
DG	7.431	24.241	44.355	127.431	417.263	353.377

The chemical composition of both FA and DG was tested by using the sequential X-ray fluorescence spectrometer from the University of Science and Technology Beijing, Beijing, China. The basic parameter including scanning velocity, voltage, and current were 300°/min, 60 kV, and 140 mA, respectively. The X-ray fluorescence spectrometry was conducted to investigate the chemical composition of tested mine specimens in this study. The cement composition of FA and DG materials were listed in Figure 2.

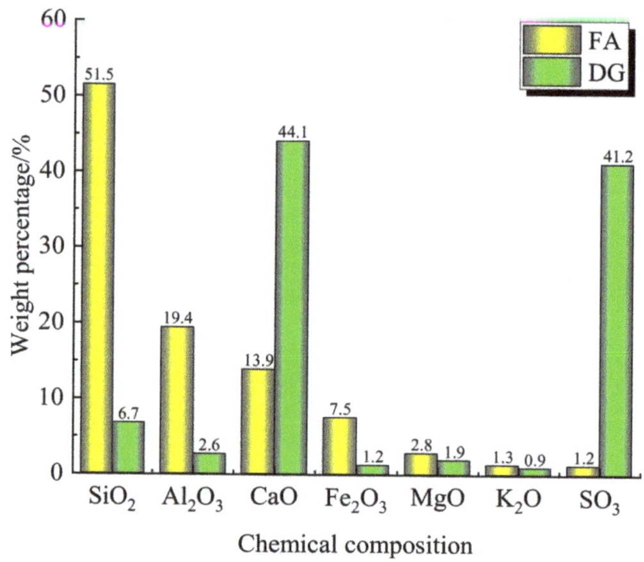

Figure 2. Chemical composition of FA and DG materials.

2.1.2. Binder and Water

In this study, the ordinary Portland cement from Chengxin Cement Manufacturing Co., Ltd from Tanshan in China named OPC 42.5R was selected as a basic binder for preparing the mine fill samples. The chemical composition of the cement used was shown in Table 2. Tap water was used to mix the FA, DG, and cement.

Table 2. Chemical composition of ordinary Portland cement (OPC) 42.5R.

Chemical Composition	SiO_2	Fe_2O_3	Al_2O_3	MgO	CaO	SO_3	K_2O
content (%)	20.1	2.91	5.11	1.57	61.8	1.98	0.37

2.2. Specimens Preparation and Curing Conditions

Firstly, an electronic scale named AUW120D 220D with an accuracy of 0.01 g from Japan was used to weigh the FA, cement, DG, and water. Secondly, dry cement, FA, and DG were mixed at least for 3 min. Tap water was poured into the stirring tank, and then the JJ-5 cement mortar mixer produced by Wuxi Jianyi Instrument Machinery Co., Ltd. was used to mix for 3 min again until the slurry was fully dispersed. The diameter and height of cylindrical molds were 50 and 100 mm, respectively. Note that a recent experimental work by Yilmaz et al. [37] has been checking specimen size effect on compressive strength of cemented tailings or paste backfills, showing that samples with a diameter of 5 and 10 cm in height always gave better strength performance than other sample sizes (D × H: 7.5 cm × 15 cm and 10 cm × 20 cm) for a given cemented tailings backfill recipe. The slurry was poured into the molds and then placed in a curing box named HSBY-40B. The temperature and humidity were set as 20 ± 1 °C and 90 ± 5%. The curing time were set as 7 and 14 days in this study. Figure 3 shows the entire process of the mine filling specimen preparation.

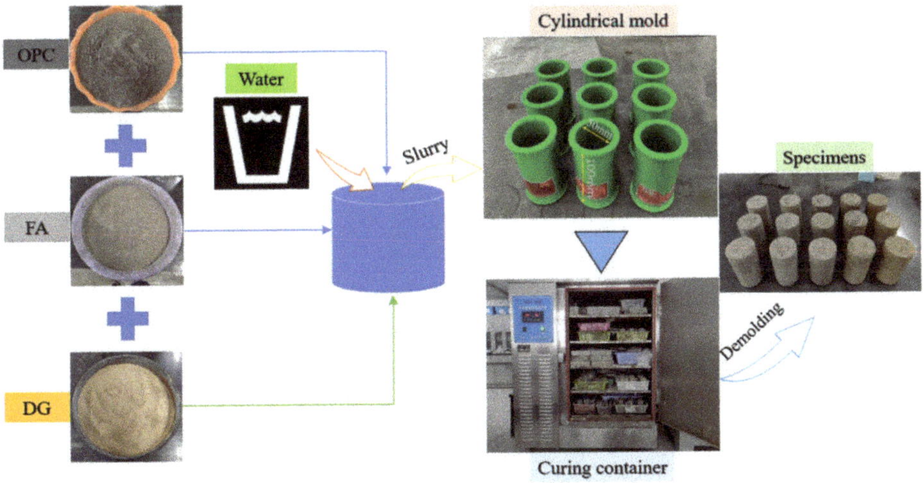

Figure 3. The preparation process of tested specimens.

In this section, the cement-to-solid waste (FA and DG) ratio was set as 1:8. Table 3 shows the proportion of each component of each sample in each group. Among them, "F-8-D-0", "F-6-D-2", "F-4-D-4", "F-2-D-6", and "F-0-D-8" indicated the number of each tested specimens, the number after F indicated the proportion of FA for each tested specimen, and the number after D indicated the proportion of DG for each tested specimen. In the five groups, each group of six specimens, a total of 30 specimens were tested.

Table 3. The proportions of each component of each specimen.

No.	OPC (wt. %)	FA (wt. %)	DG (wt. %)	Water (wt. %)
F-8-D-0	7.8	62.2	0.00	30
F-6-D-2	7.8	46.7	15.5	30
F-4-D-4	7.8	31.1	31.1	30
F-2-D-6	7.8	15.5	46.7	30
F-0-D-8	7.8	0.00	62.2	30

2.3. *Uniaxial Compressive Tests*

In this study, the both ends of the tested mine fill specimens' sections were polished to ensure flatness. As shown in Figure 4, the uniaxial compressive tests were performed to obtain their uniaxial compressive strength (UCS). The wdwi-100 loading system of the Materials Center of the University of Science and Technology Beijing (Beijing) was used for UCS testing. The loading rate was set as a fixed speed of 1 mm/min. To reduce the test errors, each group of uniaxial compression tests contained three specimens to obtain the mean UCS values for the final evaluation. The experimental data could be recorded and stored in the computer during the whole loading process.

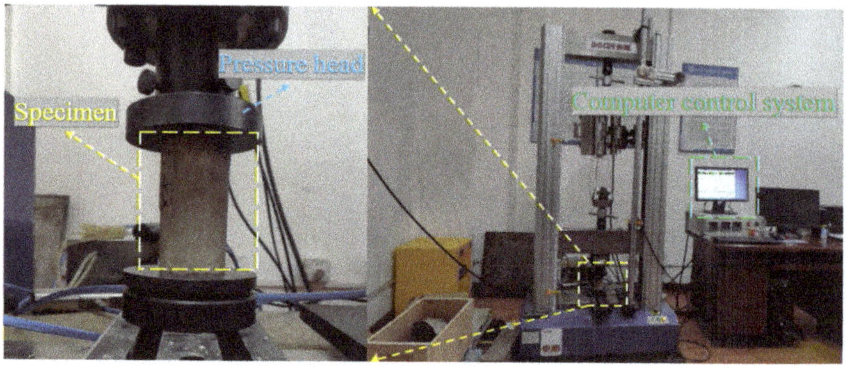

Figure 4. Photo of testing apparatus implemented in this study.

2.4. Principle of Energy Dissipation

It is known that the materials including rock, steel, and other cement-based materials deformed during the whole loading process [38] assuming that there was no heat exchanged between the physical process and the outside world. It was a closed system, and the total input energy generated by the external force was U. According to the first law of thermodynamics [39,40], we could obtain the below equations.

$$U = \int_0^{\varepsilon} \sigma d\varepsilon \quad (1)$$

$$U_e = \frac{1}{2E_0}\sigma^2 \quad (2)$$

The U_d was determined as:

$$U_\mathrm{d} = U - U_\mathrm{e} \quad (3)$$

In Equations (1)–(3), U was the total input energy, MJ·m $^{-3}$; U_d was the dissipation energy, MJ·m $^{-3}$; U_e was the strain energy that the specimens can release, MJ·m $^{-3}$; E was the unloaded elastic modulus of the samples, MPa, here was replaced by the initial elastic modulus E_0; ε was the strain of the samples; and σ was the sample's stress. Figure 5 showed the relationship between the dissipation energy and the releasable strain energy.

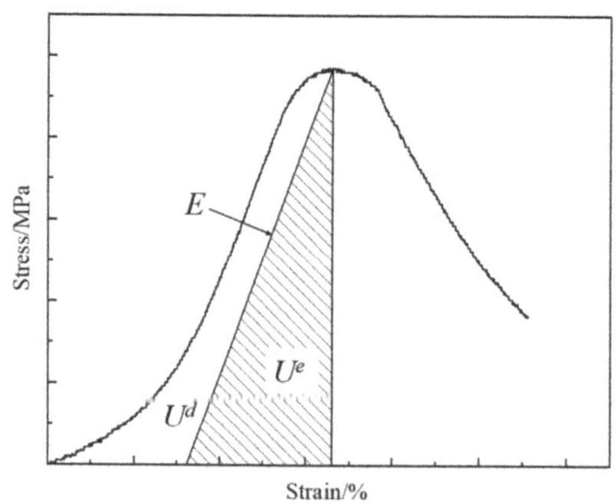

Figure 5. The relationship between U_d and U_e.

2.5. Scanning Electron Microscopy Analysis

The microstructure characteristics of the mine backfill specimens were investigated using a scanning electron microscope (SEM: Zeiss EVO18, Carl Zeiss AG, Oberkochen, Germany) system. The basic parameters of the microscope used were as follows: the accelerating voltage was between 10 and 20 kV; the maximum magnification and resolution were 2000 times and 3 nm, respectively.Before the experiment, the observed samples were dried, and then mine filling samples were sprayed with diamond by vacuum coating [41,42]. The processed samples could be observed by SEM. Figure 6 shows the SEM observation process.

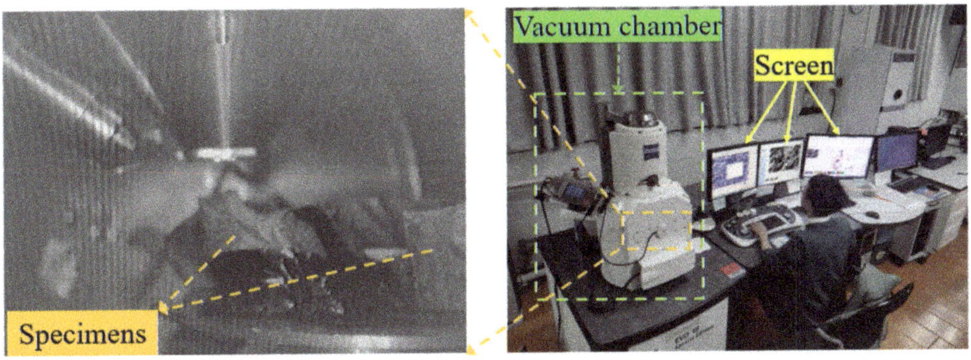

Figure 6. Photos of the used ZEISS EVO18 SEM system for microstructural investigations.

3. Results and Discussion

3.1. Effect of FA/DG Ratio on the UCS

Table 4 shows the UCS values of specimens prepared with different FA and DG proportions after curing times of 7 and 14 days. In Table 4, "1, 2, and 3" represented three specimens' number with the same ratio. It should be emphasized that the "F-6-D-2-2" specimens in the 7 and 14 day curing time did not meet the test requirements; therefore, the data were deleted. The average UCS values of mine fill samples prepared with different solid-to-waste ratios after curing time of 7 and 14 days were shown in Figure 7.

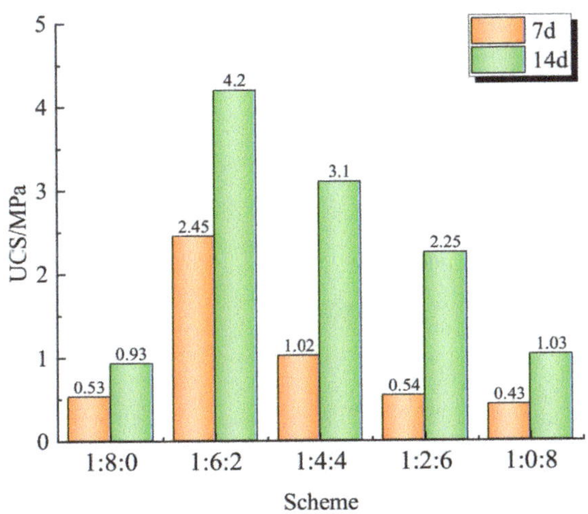

Figure 7. Average compressive strength of 7 day and 14 day cured backfill specimens.

Table 4. The uniaxial compressive strength (UCS) results of the tested mine backfill specimens.

Test No.	UCS (MPa)	Test No.	UCS (MPa)	Test No.	UCS (MPa)
		Curing Times: 7 days			
F-8-D-0-1	0.50	F-6-D-2-1	2.52	F-4-D-4-1	1.03
F-8-D-0-2	0.55	F-6-D-2-2	/	F-4-D-4-2	1.00
F-8-D-0-3	0.54	F-6-D-2-3	2.45	F-4-D-4-3	1.02
Average:	0.53	Average:	2.49	Average:	1.02
F-2-D-6-1	0.53	F-0-D-8-1	0.47		
F-2-D-6-2	0.60	F-0-D-8-2	0.38		
F-2-D-6-3	0.50	F-0-D-8-3	0.44		
Average:	0.54	Average:	0.43		
		Curing Times: 14 days			
F-8-D-0-1	0.96	F-6-D-2-1	4.35	F-4-D-4-1	3.34
F-8-D-0-2	0.83	F-6-D-2-2	/	F-4-D-4-2	3.04
F-8-D-0-3	1.00	F-6-D-2-3	4.06	F-4-D-4-3	2.91
Average:	0.93	Average:	4.20	Average:	3.10
F-2-D-6-1	2.23	F-0-D-8-1	1.07		
F-2-D-6-2	2.18	F-0-D-8-2	1.11		
F-2-D-6-3	2.34	F-0-D-8-3	0.91		
Average:	2.25	Average:	1.03		

It could be obtained that the average UCS values of the samples cured at 7 and 14 days varies with the solid-to-waste ratios. The ideal ratio was 1:6:2 (OPC:FA:DG), the average UCS values for 7 days of curing time has reached 2.49 MPa, and the average UCS value reached 4.20 MPa after curing time of 14 days. In addition, the average UCS values of the mine backfill specimens gradually decreased when the proportion of DG increased. However, the minimum average UCS value of the F-0-D-8 specimen was 0.43 MPa when the curing time was 7 days. The main reason behind this behavior was that much more DG content would affect more hydration products generated inside the tested specimens. Moreover, the UCS values of tested F-8-D-0 mine fill specimens were not so large because of the hydration reaction inside the specimens was not obvious. Accordingly, the average UCS values of tested F-6-D-2 specimens in this section could meet the UCS requirement of the cut and fill mining method.

3.2. *Effect of FA/DG Ratio on Stress–Strain Relation*

Figure 8a,b show the relationship between stress–strain of tested mine filling specimens when the curing time were 7 and 14 days, respectively.

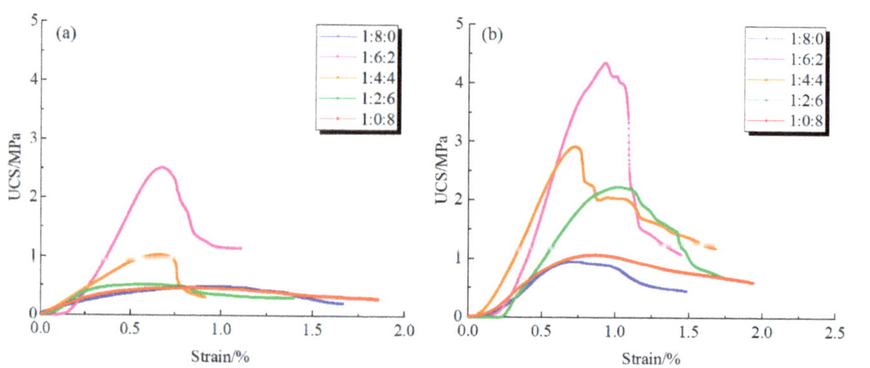

Figure 8. Stress–strain curves of the tested backfill specimens: a curing time of 7 days (**a**) and 14 days (**b**).

In general, the stress–strain curves can divide into four stages: (1) Pore compaction stage: there would be lots of pores in the tested specimens due to FA and DG particles. These pores would be compacted when the pressure increased during this stage, resulting in a concave stress–strain curve. (2) Linear elastic stage: the strain also increased with the pressure increased. Thus, the curve became a straight line. (3) Unstable rupture stage: the elastic deformation turned into plastic deformation during this stage. The UCS values gradually reached their peak with cracks growth. (4) Crack propagation stage: after the samples reached their peak, as the pressure increased, the samples' crack expanded until samples was failed. It was also found that the F-6-D-2 and F-4-D-4 tested specimens showed strong brittleness after UCS reached the peak values. However, the UCS values of F-8-D-0, F-2-D-6, and F-0-D-8 tested mine backfill specimens were not so large. The specimens showed a good ductility after reaching their UCS values.

The UCS values of the tested specimens increased with the curing time increased. The chemical hydration reaction became more adequate, and the hydration products were more produced with time. Moreover, the hydration products filled up the pores and cracks in the tested cement-based composites specimens, resulting in denser specimens. Therefore, the UCS values of tested filling specimens were improved accordingly.

3.3. Energy Dissipation Analysis

Figure 9 showed the stress–strain and energy evolution curves of the tested specimens. It was found that the input energy and the dissipation energy showed a nonlinear growth trend with the axial strain increased.

However, the elastic strain energy increased firstly and then decreased combining with the specimens' failure process during the uniaxial compression tests. It was also obtained that the energy dissipation characteristics and energy distribution laws of the specimens were different in different deformation stages. The elastic strain energy in the samples occupied a dominant position before the peak failure.

Most of the input energy was converted into elastic strain energy and stored inside samples. Samples' elastic energy slows down, while the dissipated energy increased after entering the plastic rupture stage. When the crack increased sharply, the crack accelerated and expanded. After reaching the UCS location, the cracks' growth trend further increased. The elastic energy accumulated was suddenly released and converted into the dissipation energy. At last, the tested mine backfill specimens were failed eventually.

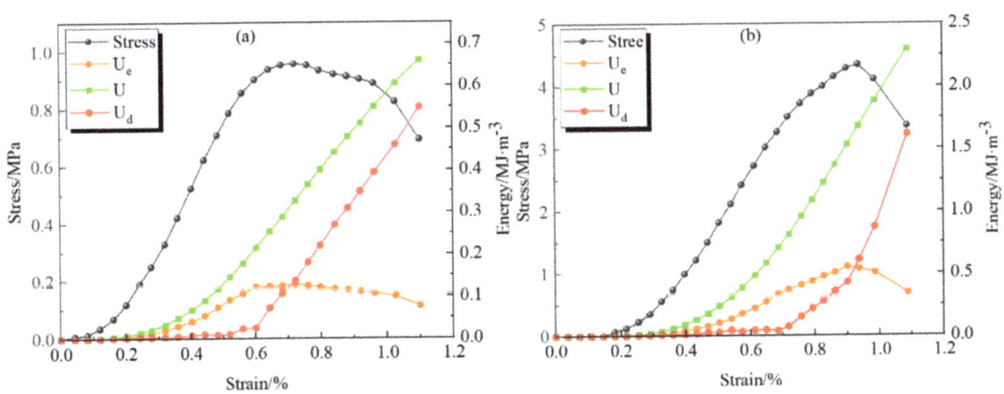

Figure 9. *Cont.*

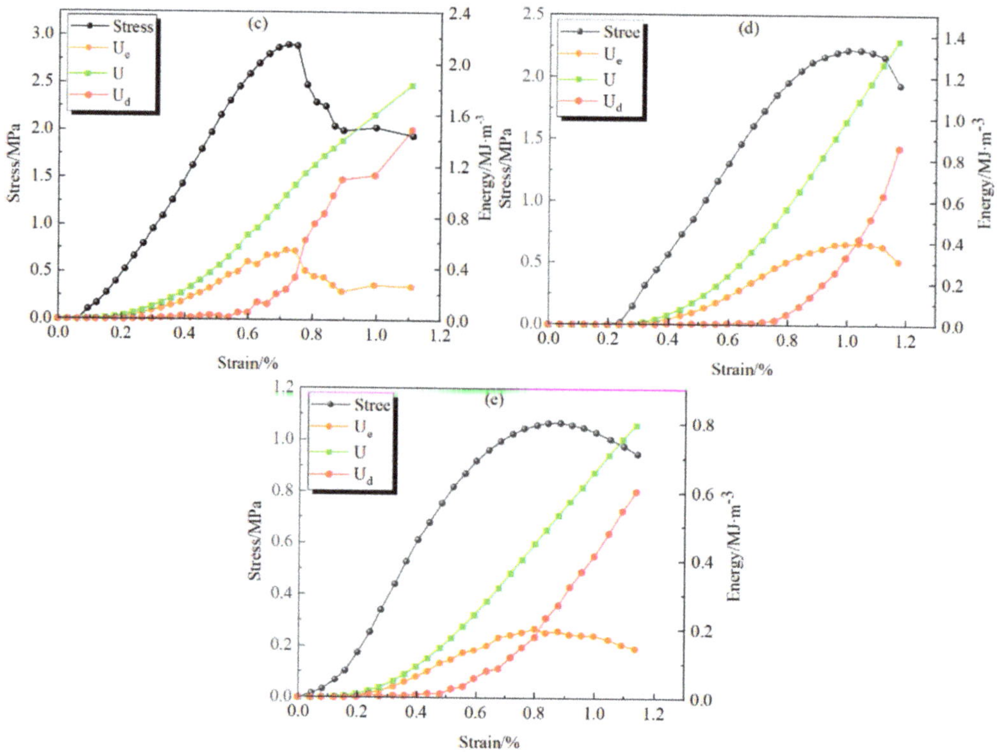

Figure 9. Samples' stress–strain and energy evolution curves: (**a**) F-8-D-0; (**b**) F-6-D-2; (**c**) F-4-D-4; (**d**) F-2-D-6; and (**e**) F-0-D-8.

3.4. The Failure Modes of BMs

Figure 10 shows the failure modes of the tested mine backfill samples with different proportions in the UCS test. It was found that the main failure modes of the F-8-D-0 and F-4-D-4 specimens were tensile failure along the loading direction. However, shear cracks mainly appeared on the surface of the F-2-D-6 and F-0-D-8 specimens. Moreover, both the shear and tensile cracks appeared on the surface of the F-6-D-2 specimen. It was also obtained that each tested mine fill sample in this study showed obvious brittleness corresponding to the stress–strain curves. With the increase of the axial loading pressure, the cement-based composites specimens showed that the accelerated crack propagation and fragments fell off the surface of the specimen.

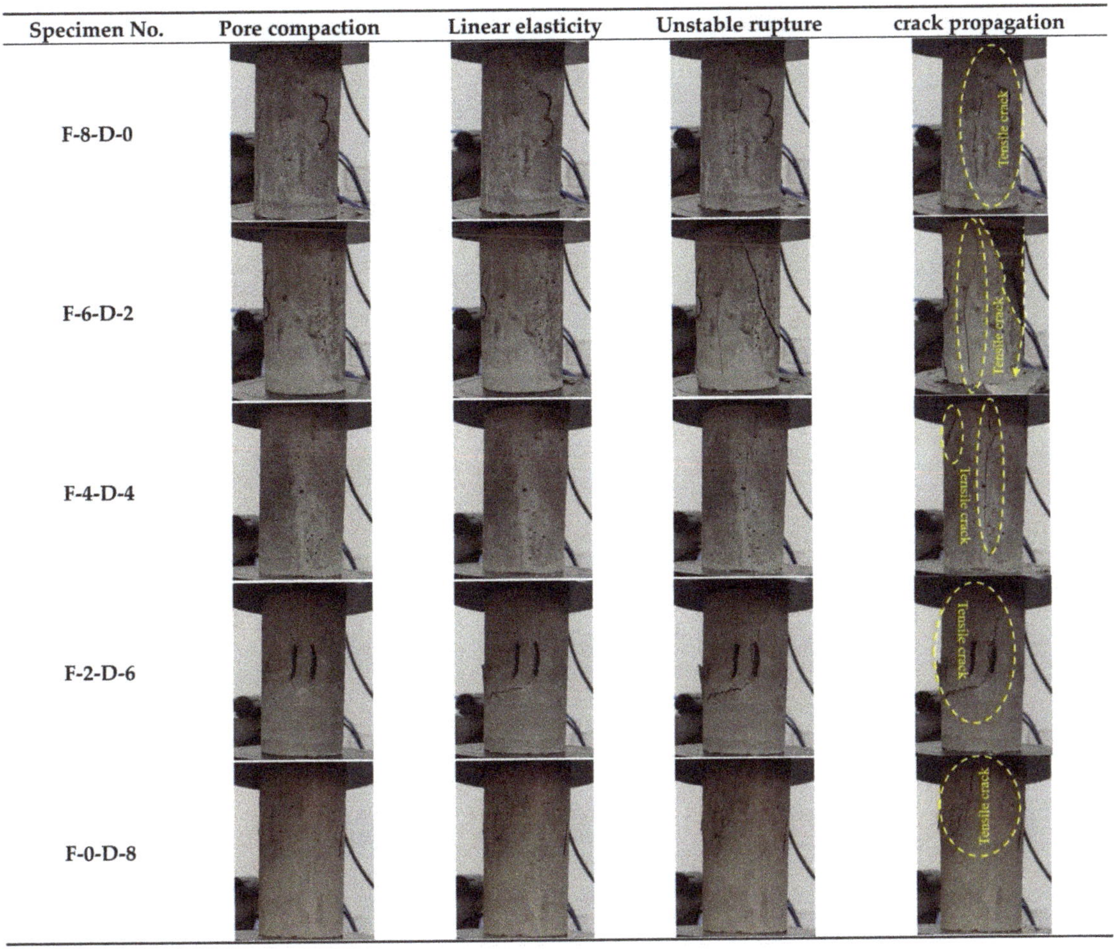

Figure 10. Failure modes of mine backfill specimens with different proportions under uniaxial compression.

3.5. Microstructural Characteristics Analysis

Figure 11 shows that SEM images of tested mine backfill samples with five different solid waste ratios when the curing time was 7 days. The ImageJ software was used to explore the microstructure of the tested backfill samples. Figure 11a shows the microstructural diagram of the F-8-D-0 samples. It was found that there were many large-size particles of FA on the sample's surface. There were also many pores on the tested specimen's surface. Moreover, the maximum particle size of FA and the largest pore size were 29.06 μm and 24.62 μm, respectively. It was also found from Figure 11b that a large amount of hydration products including calcium silicate hydrate (C-S-H), and few pores appeared on the solid-phase particles' surface. Moreover, the surface of the F-6-D-2 sample was much denser than other mine fill samples in this study. As shown in Figure 11c, lots of large-size pores existed on the F-4-D-4 sample's surface, and the maximum size reached 40.94 μm. Moreover, there were still lots of large-particle DG pieces and pores on the surface of the F-2-D-6 and F-0-D-8 mine fill samples, as demonstrated in Figure 11d,e.

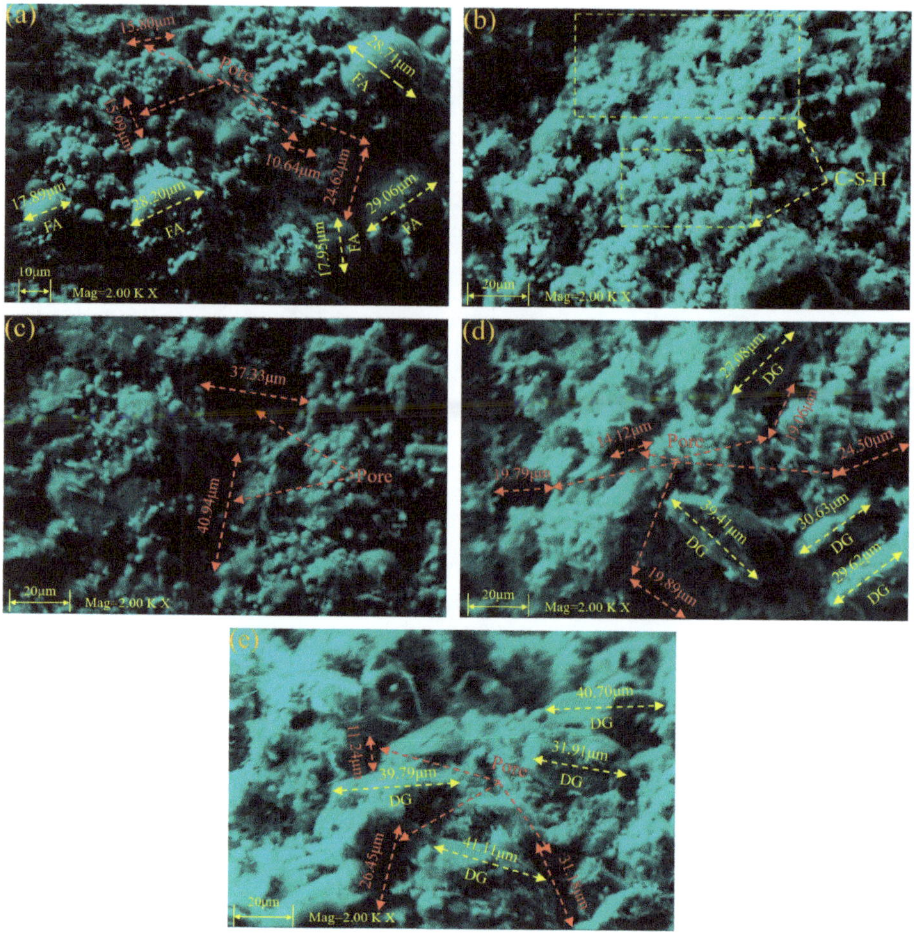

Figure 11. SEM image of 7 day cured specimens: (**a**) F-8-D-0; (**b**) F-6-D-2; (**c**) F-4-D-4; (**d**) F-2-D-6; and (**e**) F-0-D-8.

The element distribution mappings of the specimens' SEM images were shown from Figures 12–16. As shown in Figure 12, it was found that the Si, Al, and O elements were concentrated in the large amounts at the FA particles' location, which were consistent with the XRF analysis results (Figure 2). In Figure 13, the hydration reaction proceeded sufficiently and produced a large amount of flocculating substances. According to the element distribution, it was speculated that these flocculating substances were C-S-H formed by the reaction of CaO, Al_2O_3, SiO_2, and water. Moreover, the UCS values of the mine backfill specimens were strengthened because of the appearance of these C-S-H gels. Moreover, the elements including Ca, S, and O gathered in large amounts at the DG particles' locations with the proportion of DG increased, which were shown in Figures 14–16. The presence of these DG particles directly caused the UCS values of the mine backfill specimens to decrease.

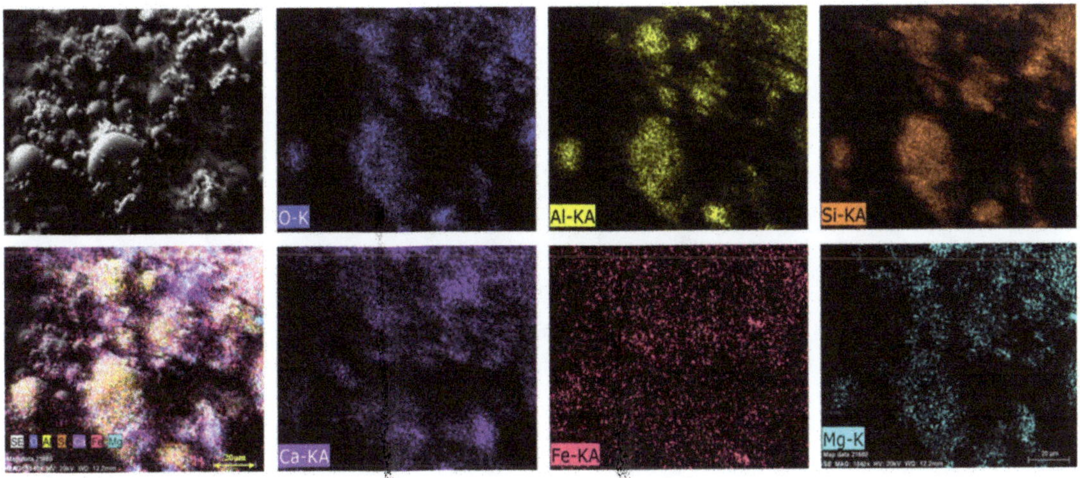

Figure 12. SEM elemental mapping of the F-8-D-0 specimen.

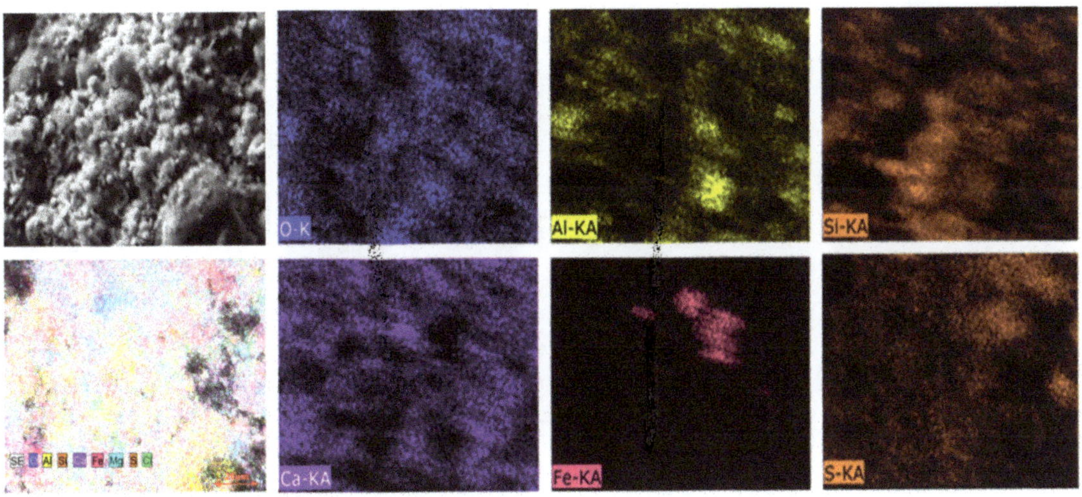

Figure 13. SEM elemental mapping of the F-6-D-2 specimen.

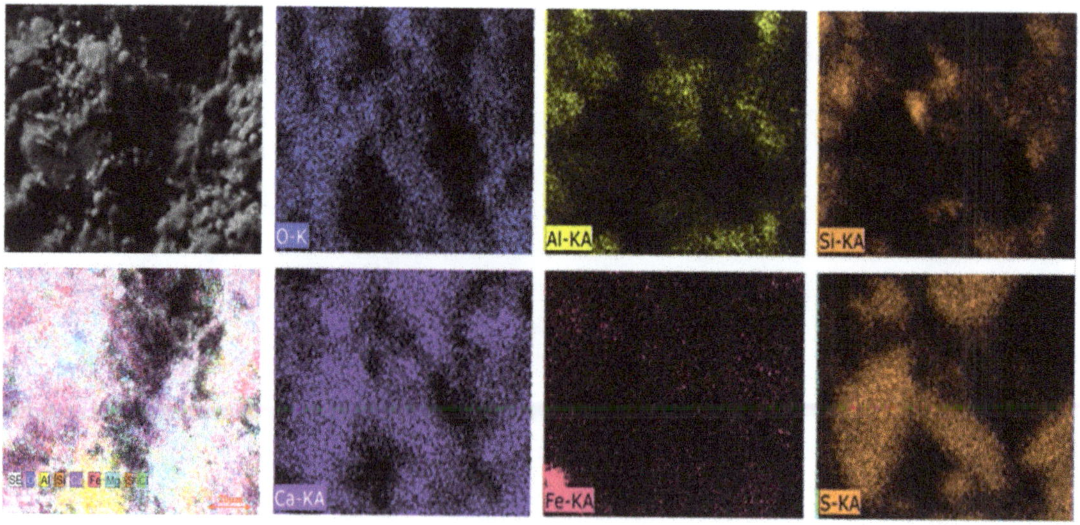

Figure 14. SEM elemental mapping of the F-4-D-4 specimen.

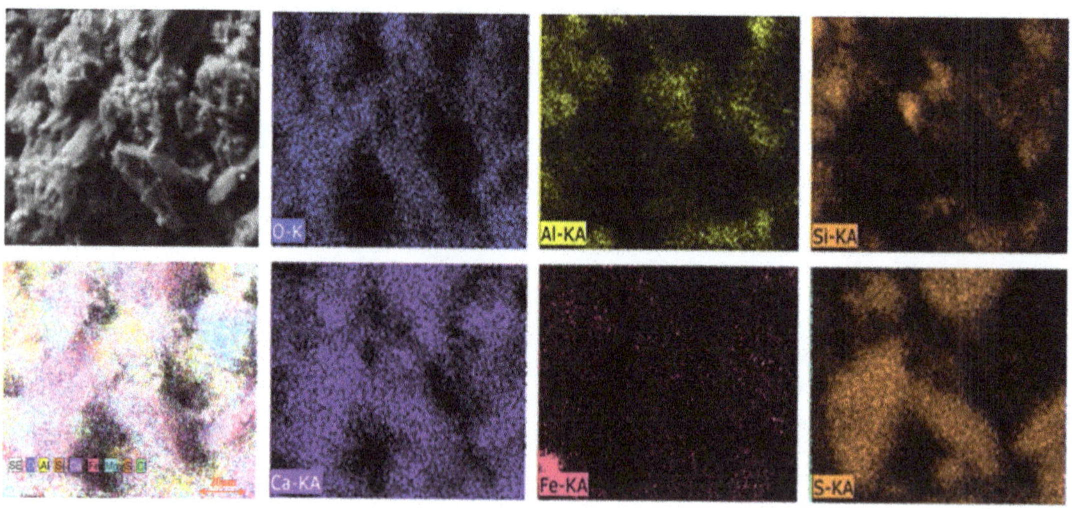

Figure 15. SEM elemental mapping of the F-2-D-6 specimen.

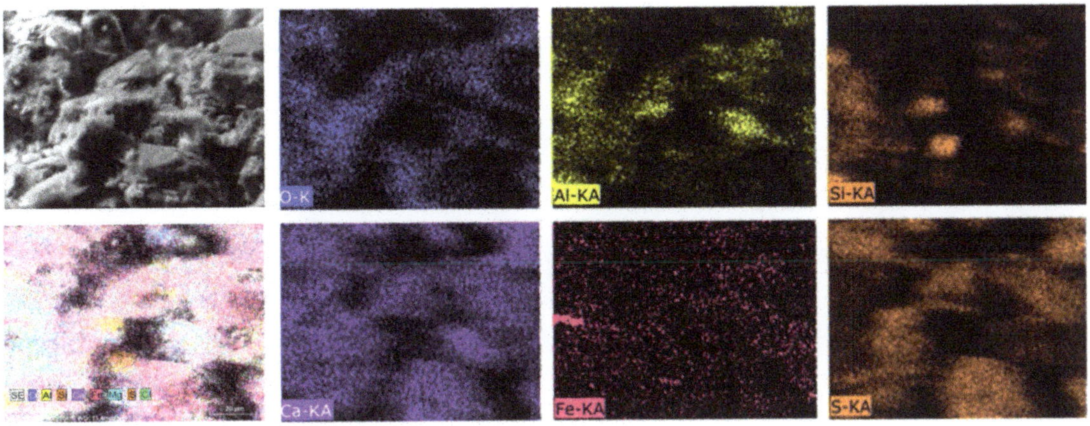

Figure 16. SEM elemental mapping of the F-0-D-8 specimen.

4. Conclusions

To explore the mechanical and microstructural characteristics of cement-based solid waste composites, some lab experiments such as uniaxial compressive strength, energy dissipation, and scanning electron microscopy were conducted. From the performed tests, the following results can be drawn:

(1) The UCS results shown that the ideal backfill mix (OPC:FA:DG/1:6:2) provided a strength gain of 2.49 and 4.20 MPa for a curing time of 7 and 14 days, respectively.
(2) The mean strengths of mine fills steadily decreased with increasing DG content. The obtained strengths meet the criteria required for the cut and fill mining system.
(3) The total input energy and elastic strain energy showed a nonlinear growth trend with increasing loads. The elastic strain energy increased first and then decreased.
(4) The dissipative energy grew gradually before samples failed. The elastic energy values accumulated in samples were suddenly released and converted into dissipative energy when samples were closed to failure.
(5) A typical backfill failure mode was tensile along the loading direction and shear cracks. Samples showed obvious brittleness corresponding to their stress–strain curves.
(6) Hydration products including calcium silicate hydrate on the surface of the ideal backfill mix were denser and had less porosity, which verified the high UCS value.

As a result, the findings of this research work have proven the feasibility of using fly ash and desulfurized gypsum to prepare backfills for underground mines. It is also good to notice that further investigations on optimization of the dosages of solid wastes used within mine backfill samples are now in progress.

Author Contributions: Conceptualization, S.C.; methodology, Z.H.; writing—original draft preparation, Z.H.; writing—review and editing and supervision, E.Y. and S.C.; funding acquisition, S.C. All authors have read and agreed to the published version of the manuscript.

Funding: This research was funded by the National Natural Science Foundation of China (Grant numbers 51804017) and the Fundamental Research Funds for Central Universities (Grant number FRF-TP-20-001A2).

Data Availability Statement: No applicable.

Conflicts of Interest: The authors declare no conflict of interest.

References

1. Zhang, W.; Zhang, Y.; Gong, J.; Yang, B.; Zhang, Z.; Wang, B.; Zhu, C.; Shi, J.; Yue, K. Comparison of the suitability of plant species for greenbelt construction based on particulate matter capture capacity, air pollution tolerance index, and antioxidant system. *Environ. Pollut.* **2020**, *263*, 114615. [CrossRef]
2. Pattanaik, A.; Mukherjee, M.; Mishra, S.; Mukharjee, M. Influence of curing condition on thermo-mechanical properties of fly ash reinforced epoxy composite. *Compos. Part B Eng.* **2019**, *176*. [CrossRef]
3. Xue, G.; Yilmaz, E.; Song, W.; Cao, S. Analysis of internal structure behavior of fiber reinforced cement-tailings matrix composites through X-ray computed tomography. *Compos. Part B Eng.* **2019**, *175*, 107091. [CrossRef]
4. Wang, D.-L.; Zhang, Q.-L.; Chen, Q.-S.; Qi, C.-C.; Feng, Y.; Xiao, C.-C. Temperature variation characteristics in flocculation settlement of tailings and its mechanism. *Int. J. Miner. Met. Mater.* **2020**, *27*, 1438–1448. [CrossRef]
5. Zhao, Y.; Taheri, A.; Karakus, M.; Chen, Z.; Deng, A. Effects of water content, water type and temperature on the rheological behaviour of slag-cement and fly ash-cement paste backfill. *Int. J. Min. Sci. Technol.* **2020**, *30*, 271–278. [CrossRef]
6. Cao, S.; Yilmaz, E.; Yin, Z.; Xue, G.; Song, W.; Sun, L. CT scanning of internal crack mechanism and strength behavior of cement-fiber-tailings matrix composites. *Cem. Concr. Compos.* **2021**, *116*, 103865. [CrossRef]
7. Wang, X.L.; Li, Q.Y.; Chen, S.C.; Yue, G.B. Application research and prospect of industrial waste in new building materials. *Bull. Chin. Ceram. Soc.* **2019**, *38*, 3456–3464.
8. Yilmaz, E.; Belem, T.; Benzaazoua, M.; Kesimal, A.; Ercikdi, B.; Cihangir, F. Use of high-density paste backfill for safe disposal of copper/zinc mine tailings. *Miner. Resour. Manag.* **2011**, *27*, 81–94.
9. Glosser, D.; Suraneni, P.; Isgor, O.B.; Weiss, W.J. Using glass content to determine the reactivity of fly ash for thermodynamic calculations. *Cem. Concr. Compos.* **2021**, *115*, 103849. [CrossRef]
10. Hansen, S.; Sadeghian, P. Recycled gypsum powder from waste drywalls combined with fly ash for partial cement replacement in concrete. *J. Clean. Prod.* **2020**, *274*, 122785. [CrossRef]
11. Jia, R.; Wang, Q.; Feng, P. A comprehensive overview of fibre-reinforced gypsum-based composites (FRGCs) in the construction field. *Compos. Part B Eng.* **2021**, *205*, 108540. [CrossRef]
12. Thymotie, A.; Chang, T.-P.; Nguyen, H.-A. Improving properties of high-volume fly ash cement paste blended with β-hemihydrate from flue gas desulfurization gypsum. *Constr. Build. Mater.* **2020**, *261*, 120494. [CrossRef]
13. Li, Y.; Liu, X.; Li, Z.; Ren, Y.; Wang, Y.; Zhang, W. Preparation, characterization and application of red mud, fly ash and desulfurized gypsum based eco-friendly road base materials. *J. Clean. Prod.* **2021**, *284*, 124777. [CrossRef]
14. Yan, B.; Lai, X.; Jia, H.; Yilmaz, E.; Hou, C. A Solution to the Time-Dependent Stress Distribution in Suborbicular Backfilled Stope Interaction with Creeping Rock. *Adv. Civ. Eng.* **2021**, *2021*, 5533980. [CrossRef]
15. Fořt, J.; Šál, J.; Ševčík, R.; Doleželováa, M.; Keppert, M.; Jerman, M.; Záleská, M.; Stehel, V.; Černýa, R. Biomass fly ash as an alternative to coal fly ash in blended cements: Functional aspects. *Constr. Build. Mater.* **2020**, *271*, 121544. [CrossRef]
16. Gu, K.; Chen, B. Research on the incorporation of untreated flue gas desulfurization gypsum into magnesium oxysulfate cement. *J. Clean. Prod.* **2020**, *271*, 122497. [CrossRef]
17. Fu, B.; Liu, G.; Mian, M.; Sun, M.; Wu, D. Characteristics and speciation of heavy metals in fly ash and FGD gypsum from Chinese coal-fired power plants. *Fuel* **2019**, *251*, 593–602. [CrossRef]
18. Cavusoglu, I.; Yilmaz, E.; Yilmaz, A.O. Sodium silicate effect on setting properties, strength behavior and microstructure of cemented coal fly ash backfill. *Powder Technol.* **2021**, *384*, 17–28. [CrossRef]
19. Wang, T.; Wu, K.; Wu, M. Development of green binder systems based on flue gas desulfurization gypsum and fly ash incorporating slag or steel slag powders. *Constr. Build. Mater.* **2020**, *265*, 120275. [CrossRef]
20. Yang, L.; Jing, M.; Lu, L.; Zhu, X.; Zhao, P.; Chen, M.; Li, L.; Liu, J. Effects of modified materials prepared from wastes on the performance of flue gas desulfurization gypsum-based composite wall materials. *Constr. Build. Mater.* **2020**, *257*, 119519. [CrossRef]
21. Neto, H.F.D.S.; Pereira, W.V.D.S.; Dias, Y.N.; de Souza, E.S.; Teixeira, R.A.; de Lima, M.W.; Ramos, S.J.; Amarante, C.B.D.; Fernandes, A.R. Environmental and human health risks of arsenic in gold mining areas in the eastern Amazon. *Environ. Pollut.* **2020**, *265*, 114969. [CrossRef]
22. Li, J.; Yilmaz, E.; Cao, S. Influence of Solid Content, Cement/Tailings Ratio, and Curing Time on Rheology and Strength of Cemented Tailings Backfill. *Minerals* **2020**, *10*, 922. [CrossRef]
23. Gao, S.; Cui, X.; Kang, S.; Ding, Y. Sustainable applications for utilizing molybdenum tailings in concrete. *J. Clean. Prod.* **2020**, *266*, 122020. [CrossRef]
24. Cao, S.; Xue, G.; Song, W.; Teng, Q. Strain rate effect on dynamic mechanical properties and microstructure of cemented tailings composites. *Constr. Build. Mater.* **2020**, *247*, 118537. [CrossRef]
25. Cao, S.; Xue, G.; Yilmaz, E.; Yin, Z.; Yang, F. Utilizing concrete pillars as an environmental mining practice in underground mines. *J. Clean. Prod.* **2021**, *278*, 123433. [CrossRef]
26. Xue, G.; Yilmaz, E.; Feng, G.; Cao, S.; Sun, L. Reinforcement effect of polypropylene fiber on dynamic properties of cemented tailings backfill under SHPB impact loading. *Constr. Build. Mater.* **2021**, *279*, 122417. [CrossRef]
27. Xue, G.; Yilmaz, E.; Song, W.; Cao, S. Fiber length effect on strength properties of polypropylene fiber reinforced cemented tailings backfill specimens with different sizes. *Constr. Build. Mater.* **2020**, *241*, 118113. [CrossRef]

28. Cao, S.; Zheng, D.; Yilmaz, E.; Yin, Z.; Xue, G.; Yang, F. Strength development and microstructure characteristics of artificial concrete pillar considering fiber type and content effects. *Constr. Build. Mater.* **2020**, *256*, 119408. [CrossRef]
29. Chen, Q.; Sun, S.; Liu, Y.; Qi, C.; Zhou, H.; Zhang, Q. Experimental and numerical study on immobilization and leaching characteristics of fluoride from phosphogypsum based cemented paste backfill. *Int. J. Miner. Metall. Mater.* **2021**, in press. [CrossRef]
30. Jiang, H.; Yi, H.; Yilmaz, E.; Liu, S.; Qiu, J. Ultrasonic evaluation of strength properties of cemented paste backfill: Effects of mineral admixture and curing temperature. *Ultrasonics* **2020**, *100*, 105983. [CrossRef]
31. Marieta, C.; Guerrero, A.; Leon, I. Municipal solid waste incineration fly ash to produce eco-friendly binders for sustainable building construction. *Waste Manag.* **2021**, *120*, 114–124. [CrossRef]
32. Cao, S.; Xue, G.; Yilmaz, E. Flexural Behavior of Fiber Reinforced Cemented Tailings Backfill under Three-Point Bending. *IEEE Access* **2019**, *7*, 139317–139328. [CrossRef]
33. Wansom, S.; Chintasongkro, P.; Srijampan, W. Water resistant blended cements containing flue-gas desulfurization gypsum, Portland cement and fly ash for structural applications. *Cem. Concr. Compos.* **2019**, *103*, 134–148. [CrossRef]
34. Navarrete, I.; Vargas, F.; Martinez, P.; Paul, A.; Lopez, M. Flue gas desulfurization (FGD) fly ash as a sustainable, safe alternative for cement-based materials. *J. Clean. Prod.* **2021**, *283*, 124646. [CrossRef]
35. Li, J.; Zhang, S.; Wang, Q.; Ni, W.; Li, K.; Fu, P.; Hu, W.; Li, Z. Feasibility of using fly ash–slag-based binder for mine backfilling and its associated leaching risks. *J. Hazard. Mater.* **2020**, *400*, 123191. [CrossRef] [PubMed]
36. Liu, C.; Wang, F.; Zhang, M. Modelling of 3D microstructure and effective diffusivity of fly ash blended cement paste. *Cem. Concr. Compos.* **2020**, *110*, 103586. [CrossRef]
37. Yilmaz, E.; Belem, T.; Benzaazoua, M. Specimen size effect on strength behavior of cemented paste backfills subjected to different placement conditions. *Eng. Geol.* **2015**, *185*, 52–62. [CrossRef]
38. Li, P.; Cai, M.F.; Wang, P.T.; Guo, Q.F.; Miao, S.J.; Ren, F.H. Mechanical properties and energy evolution of jointed rock specimens containing an opening under uniaxial loading. *Int. J. Miner. Metall. Mater.* **2020**. [CrossRef]
39. Cao, S.; Yilmaz, E.; Song, W.; Xue, G. Assessment of Acoustic Emission and Triaxial Mechanical Properties of Rock-Cemented Tailings Matrix Composites. *Adv. Mater. Sci. Eng.* **2019**, *2019*, 6742392. [CrossRef]
40. Han, Z.Y.; Li, D.Y.; Zhu, Q.Q.; Liu, M.; Li, X.B. Uniaxial compression failure and energy dissipation of marble specimens with flaws at the end surface. *Chin. J. Eng.* **2020**, *12*, 1–10.
41. Hefni, M.; Hassani, F. Experimental development of a novel mine backfill material: Foam mine fill. *Minerals* **2020**, *10*, 564. [CrossRef]
42. Cao, S.; Yilmaz, E.; Song, W. Evaluation of viscosity, strength and microstructural properties of cemented tailings backfill. *Minerals* **2018**, *8*, 352. [CrossRef]

minerals

MDPI

Article

Safety Analysis of Synergetic Operation of Backfilling the Open Pit Using Tailings and Excavating the Ore Deposit Underground

Qinli Zhang, Bingyi Zhang, Qiusong Chen *, Daolin Wang and Xiang Gao

School of Resources and Safety Engineering, Central South University, Changsha 410083, China; zhangqinlicn@126.com (Q.Z.); zhangbingyicsu@163.com (B.Z.); daolinw@csu.edu.cn (D.W.); xiang_gao4@Shimaogroup.com (X.G.)
* Correspondence: qiusong.chen@csu.edu.cn; Tel.: +86-151-1627-9873

Abstract: The transition from open pit mining to underground mining is essential for mineral resources to achieve deep excavation. Recently, cemented paste backfill (CPB) has been proposed as a novel technology to achieve open pit backfill (OPB). The proposed method not only eliminates the danger of the open-pit slope but also reduces the disposal of waste tailings. In order to ensure safe mining during the synergetic operation of OPB and underground mining, it is of great significance to improve this technology. In the present study, an open-pit metal mine in Anhui Province was taken as the research object. Then, the safety of underground stope roofs, underground backfill pillars, and open-pit slopes was evaluated during OPB. To this end, numerical simulations were performed and experiments were conducted on a similar physical model. Accordingly, the backfill mechanical parameters were optimized. The obtained results show that backfill height exerts the most significant effect on the safety of roofs and underground backfill pillars, accompanied by small displacements along the vertical direction during the backfill process. Moreover, concentration was observed at the foot of the slope, while the overall structure remained stable with no considerable displacement. The overall safety factors met the safety requirements. Based on the obtained results, the optimal foundation strength, foundation height, backfill strength and backfill height were 4 MPa, 10 m, 1.5 MPa, and 120 m, respectively. Moreover, it was concluded that displacements in the abovementioned three regions tend to be stable when the backfill height exceeds 150 m without damage. The present article provides a certain theoretical and application guideline for OPB practices in similar metal mines and suggests possibilities for cleaner production.

Keywords: cemented paste backfill; backfill strength; transition from open pit to underground mining; tailings disposal; numerical simulation

Citation: Zhang, Q.; Zhang, B.; Chen, Q.; Wang, D.; Gao, X. Safety Analysis of Synergetic Operation of Backfilling the Open Pit Using Tailings and Excavating the Ore Deposit Underground. *Minerals* **2021**, *11*, 818. https://doi.org/10.3390/min11080818

Academic Editor: Abbas Taheri

Received: 7 July 2021
Accepted: 26 July 2021
Published: 28 July 2021

1. Introduction

Considering reasonable economic parameters (such as stripping ratio) in the mining industry, it is crucial to transition from the open pit to the underground mining after reaching the 'great depth' in open-pit metal mines [1]. Such conversion can make full use of existing infrastructures and equipment to facilitate the excavation of ore deposits, which is of great significance for the sustainable development of metal mines. Further investigations reveal that the open pit caused by the open-pit mining not only threatens the local environment, but the created puddle affects the safety of the underground mining. On the other hand, it increases occurrence probability of collapses and landslides (Figure 1a), thereby becoming a major source of safety hazards during the normal operation of metal mines [2].

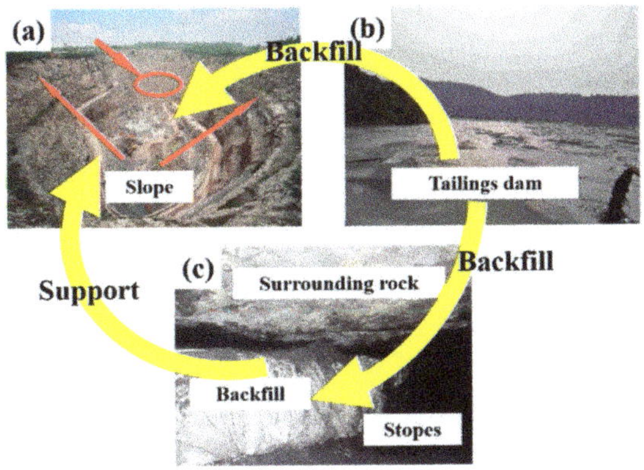

Figure 1. Relationship between slopes (**a**), tailings dam (**b**), and stopes (**c**) in surface and underground mining.

In the past few decades, increasingly large quantities of mine tailings have been produced as a result of excavating mineral resources [3,4]. Furthermore, more than 120,000 tailing dams have been established worldwide to deposit about 25 billion tons of untreated mine tailings (Figure 1b), resulting in the contamination of large amounts of Earth's surface and water, respectively [5]. A tailing dam, as a place to store mine tailings, has many drawbacks, such that establishing new tailing dams is not allowed in many countries. On the other hand, the comprehensive utilization rate of the mine tailing in China is less than 30% [6,7]. Consequently, it is necessary to increase the disposal efficiency of mine tailings. Recently, cemented paste backfill (CPB) has been proposed as an innovative technology to fill mined-out voids (also known as stopes) and effectively reduce environmental damages originating from tailing piles on the ground surface (Figure 1c). The scheme, which consists of a binder (e.g., cement), aggregates (e.g., mine tailings), and mixing water [8] can be effectively applied to eliminate hazards in underground stopes and improve the safety of underground mining operations, especially the recovery of ore bodies. Figure 1 illustrates the relationship between slopes, tailings dam, and stopes.

CPB is the most eco-friendly method to resolve tailings problems [9]. Therefore, it should be adopted to fill the open pit. It cannot only solve challenges originating from the disposal of mine tailings, but can also provide conditions for subsequent ecological restorations in open pits. Reviewing the literature on the subject indicates that the CPB method has been successfully conducted in some metal mines under different conditions. For instance, Lu et al. (2018) applied tailings to backfill the open pit of Shirengou Iron Mine (SIM), which realized 100% utilization of waste tailings and achieved clean production for its mining. Additionally, a method was proposed to combine the desulphurization technology and CPB, which can improve the disposal of waste tailings and support for surrounding rocks [10]. Simultaneously, an open pit mine planning stochastic integer program with tailings disposal was established and applied at an iron ore deposit located in Labrador, Canada, which can greatly contribute to ore production and disposal of tailings [6]. Nevertheless, there is no comprehensive specification or standard to implement with respect to OPB, which leads to a lack of evidence on the influence of its practice on slopes, roofs and underground backfill pillars when OPB is processing. In order to resolve this shortcoming, this issue is considered as the research object of the present study.

In this article, an open pit in Anhui was considered as the case study. It has a length of 1250 m, a width of 500 m, a volume of about 43 million cubic meters. Its highest slope has a height of 482 m. In order to accomplish rational disposal and ecological restoration of the open pit, metal mine tailings were used for OPB. Since the metal mine is in the transition stage from open pit to underground mining, safety issues have become the key factor in the implementation of this project. Therefore, this research aims to analyze the safety of synergetic operation of backfilling the open pit using tailings and excavating the ore deposit underground, including the stability of slopes, roofs, and underground backfills. Noted that the study does not take into account the possibility of a hazardous impact of CPB on groundwater due to toxic components in tailings. The present article is expected to provide a basis for OPB and eco-friendly production of metal mineral resources.

2. Methods and Proposals

2.1. Study Site

The present study is focused on an open-pit metal mine from the open pit to the underground mining in Anhui province, China. The metal mine is located in the middle section of the copper-iron metallogenic belt in the middle and lower reaches of the Yangtze River. The geotectonic structure belongs to the Ma'anshan-Guichi fault fold belt of the Lower Yangtze depression fold belt. The main strata exposed in the mining area are Silurian, Upper Devonian, Carboniferous, Permian, and Quaternary. The strata in the eastern part of the mining area strikes northeast and tends to northwest, with an inclination of 45° or more in the upper part and about 20° in the lower part.

In the studied mine, a two-step mining method was utilized. The first step consists of a 15 m wide room, followed by a 10 m pillar in the second step, which perceives excavated pillars from bottom to top in a layered form. Then, CPB is filled to create a platform, allowing upward mining operation. Figure 2 presents the schematic diagram of the studied open pit. When pillars are mined and filled by CPB, the same process is performed to excavate the CPB-filled room. It should be noted that the ore bodies to be excavated are below the −156 m level. In particular, the ore bodies in the middle section of −380 m are mined first, and then the middle section of −320 m, −270 m, and −230 m is mined, respectively, leaving a boundary pillar of 24 m high between the −180 m and −156 m levels.

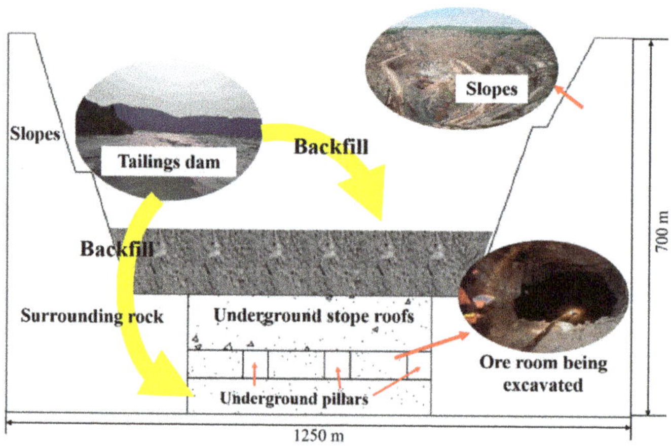

Figure 2. Open pit and underground stopes.

Due to the open pit and high-steep slope caused by the open-pit mining shown in Figure 2, a great risk threatens the entire excavation process of lower ore bodies. This process may also affect the stability of the slope [11]. Accordingly, it is necessary to consider the stability of underground stopes (including roofs and underground backfill pillars) and the slope to prevent geological disasters. In engineering practices, tailings, cement and mixing water in a certain proportion are transported to the filling slurry preparation station through pipelines to prepare homogeneous slurry by utilizing mixing equipment, which will be transported to underground stopes and the open-pit through the pipeline system.

2.2. Numerical Simulation Tests

Among different numerical methods, explicit finite volume analysis has the advantages of its simple process, which can reflect dynamic parameters such as stress and strain [12]. In the present study, post-processing and explicit finite volume analysis of Flac3D 6.0 software are employed to simulate and analyze the stability of the roofs, underground filling column, and slopes during the underground mining at different backfilling conditions. Moreover, the elastoplastic material model and the Mohr-Coulomb strength criterion were used to evaluate the stability [13,14].

The principal stress, which is the main parameter in rock mechanics, can be expressed in the form below [15,16]:

$$\sigma_3 = \sigma_1 \tan^2(45^\circ - \varphi/2) - 2c \cdot \tan(45^\circ - \varphi/2) \tag{1}$$

$$f = \frac{\sigma_3}{\sigma_3'} \tag{2}$$

where σ_1 is the maximum principal stress (compressive stress is negative and tensile stress is positive), MPa; σ_3 is the theoretical minimum principal stress, MPa; σ_3' is the calculated minimum principal stress, MPa; φ denotes the internal friction angle of the material, °; and f is a safety factor. When $f < 1$, damage will occur.

Boundary conditions conducted in this model are as follows. In the numerical simulation calculation, the horizontal displacement of the left and right boundaries of the model was fixed, the displacement at the bottom of the model was fixed, and the top of the model was set as a free surface. In terms of boundary stress, trapezoidal stress was applied to the left and right of the model. The effects of boundary conditions on the results of the numerical simulation can be ignored.

2.2.1. Modeling

A numerical model of OPB was established in the Flac3D environment. Figure 3 indicates that dimensions of the model along X-, Y- and Z-axes are 1250 m, 500 m, and 700 m, respectively. It is worth noting that an ideal elastoplastic body was selected as the ore rock mass to simplify the analysis, which ignored slight differences between ore rock mass components to simplify the simulation process. In this regard, the backfills are assumed to be in close contact with the surrounding rock and fully connected to the top after stopes are filled. In other words, the effect of "saw-tooth" reinforcement of the backfill on the slope steps is ignored. Moreover, two test schemes with different total backfill heights were subjected to the test to reflect the safety effect on each region during the OPB operation.

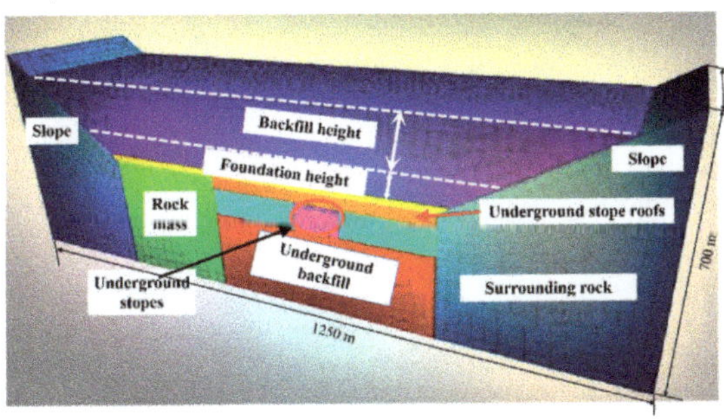

Figure 3. Numerical model.

2.2.2. Parameters

Considering the existing geological data of the studied mine and experiments of CPB, the mechanical parameters of the ore rock and underground backfills are summarized in Table 1. Furthermore, Supplementary Table S1 presents the backfill parameters of the open pit.

Table 1. Mechanical parameters of underground mining rock and underground backfill.

Materials	Elastic Modulus (GPa)	Poisson's Ratio (μ)	Density (t/m³)	Compressive Strength (MPa)	Tensile Strength (MPa)	Cohesion (MPa)	Internal Friction Angle (°)
Granite	33.81	0.26	2.73	50.38	4.16	6.23	41.20
Sandstone	8.33	0.26	2.71	17.60	7.05	4.54	34.50
Pyrite	10.20	0.18	3.83	13.20	1.57	3.68	40.70
CPB*	0.185	0.28	1.92	2.20	0.28	0.31	41.91

Note: Granite, sandstone, pyrite and CPB* represent the slope rock, underground rock, ore body, and underground backfill, respectively.

2.2.3. Simulation Schemes

In the present article, a two-step mining process was applied to analyze the stability of the abovementioned three regions with the highest risks during underground mining [17]. The top pressure corresponding to the applied parameters was calculated. Based on the engineering experience, four factors, including the foundation height, foundation strength (28 d), backfill height, and backfill strength (28 d) were selected in calculations through orthogonal experiments. This foundation represents a high-strength backfill filled at the bottom of the open pit. 'Backfill' of 'backfill strength and height' in the factors selected is denoted as backfills above the foundation, which has lower strength compared with the foundation [18]. Based on the construction guideline for reinforced cemented tailings backfill, each parameter was divided into five levels (as shown in Table 2).

Table 2. Backfill parameter level.

Level	Foundation Strength (MPa)	Foundation Height (m)	Backfill Strength (MPa)	Backfill Height (m)
1	2.0	4	0.5	100
2	2.5	6	1.0	120
3	3.0	8	1.5	140
4	3.5	10	2.0	160
5	4.0	12	2.5	180

2.3. *Physical Model*

In order to investigate the safety of the roofs, underground backfill pillars and the slopes during the backfilling process of the open pit, a physical model was prepared according to the procedure proposed by previous research [19,20]. Moreover, an orthogonal test method was adopted to design a test scheme for similar materials, and investigate stress and displacement at the roof, underground backfill, and the slope under different backfill parameters [21].

2.3.1. Similar Materials

The selection of similar materials plays a crucial role in the physical model. Based on the similarity principles [22,23] and characteristics of materials, barite powder and quartz sand were chosen as aggregates. Cement (ordinary Portland cement 425#), gypsum, and sawdust were used as binders and an additive to conduct similar material tests (Figure 4). The five materials mentioned above were all obtained from a material building factory in Changsha. Meanwhile, C_ρ, C_g, C_θ, C_E, and C_σ were defined the as similarity ratio of the density, acceleration of the gravity, internal friction angle, elastic modulus, and stress, respectively. These values are summarized in Table 3.

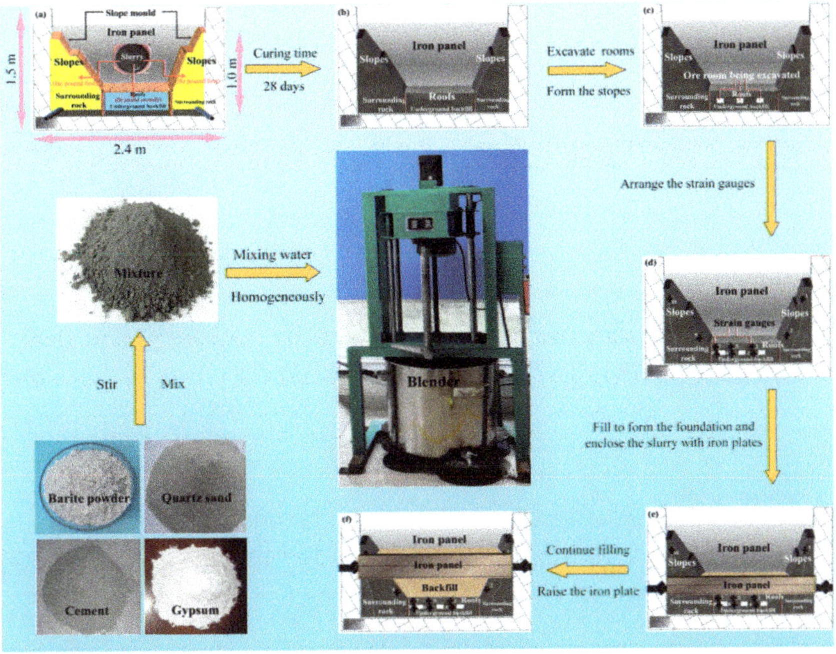

Figure 4. Schematic diagram of the modeling process (**a**–**f**).

Table 3. Similarity ratios.

Similarity Ratio	Value
C_ρ	1
C_g	1
C_θ	1
C_E	190
C_σ	190

2.3.2. Parameters

The ideal mechanical parameters of each part of the model can be determined by the similarity principle. Supplementary Table S1 presents the relevant mechanical parameters. The uniform test was performed to select proper similar materials [24]. In this regard, four factors, including the masses of the barite powder, sand, gypsum and cement are considered. Meanwhile, 12 levels and corresponding results of uniform tests are considered and shown in Supplemenary Tables S2 and S3, respectively. The main objective of these levels is to control similarity ratios, thereby determining the optimal plan accurately and minimizing the calculation error. The calculations were conducted with different mechanical parameters, including density, elastic modulus, compressive strength, internal friction angle and adhesion force [19]. Then, the obtained results were compared with ideal mechanical parameters. Accordingly, appropriate parameters and the corresponding similarity ratios of the model components were selected as shown in Table 4.

Table 4. Comparison of mechanical parameters of surrounding rock, ore body, underground backfill prototype materials, and similar materials.

Region	Parameters	Prototype Material	Similar Material	Actual Similarity Ratio	Theoretical Similarity Ratio
Surrounding rock	Cohesion (MPa)	6.23	0.038	163.95	190
	Internal friction angle (°)	41.2	41.2	1.00	1
	Compressive strength (MPa)	50.38	0.28	179.93	190
	Density (kg/m^3)	2730	2190	1.25	1
Ore body	Cohesion (MPa)	3.68	0.02	184.00	190
	Internal friction angle (°)	40.7	37.6	1.08	1
	Compressive strength (MPa)	13.2	0.1	132.00	190
	Density (kg/m^3)	3830	2740	1.40	1
Underground backfill	Cohesion (MPa)	0.31	-	-	190
	Internal friction angle (°)	41.91	37.7	1.11	1
	Compressive strength (MPa)	2.20	-	-	190
	Density (kg/m^3)	1920	2180	0.88	1

It should be noted that the effect caused by a difference between theoretical and actual similarity ratios on tests results can be regarded as negligible.

2.3.3. Modeling Process

Modeling the Open Pit

In this section, the rock mass plane similarity model test system (ZYDL-YS120) was used to establish the physical model shown in Figure 4a [20]. In the present study, the model height and length of the slope were set to 1.0 and 2.0 m, respectively. It can be observed that the densities of similar materials of surrounding rock (2190 kg/m^3) and underground backfill (2180 kg/m^3) are similar in value. Based on the parameters of the selected proportions (Supplementary Table S3) and results obtained from numerical simulations, mixtures of schemes 8, 12, and 8 were chosen as materials of the surrounding rock, ore body and underground backfill, respectively. As can be seen in Figure 4, materials were mixed and stirred for 15–30 s by the blender to prepare a homogeneous slurry with a water-to-binder ratio of 0.5 in accordance with Chinese national standard GB/T 12957-2005 and the reference test [25]. Several wooden boards were combined as molds to construct the slope, as shown in Figure 4a. The slurry was first poured into the slopes regions on both sides. After the slurry became hardened, the region in the middle was filled. After 28 days of curing in an environment with a temperature of 25 $\pm$ 2 °C and a humidity of 90%, molds were removed (Figure 4b) and the ore rooms in the underground stope were excavated using a concrete cutter with diamond saw blades (500-type) to provide in-situ conditions for underground mining (Figure 4c). Three regions, including the roofs, underground backfill pillars, and slopes, were arranged as main monitoring points to determine the strain development during the OPB operation. Then two strain gauges were installed horizontally and vertically at 10 points to reflect the horizontal and vertical strain at these measuring points (Figure 4d) [26].

OPB

Based on the obtained results from the numerical simulations, which will be discussed in Section 3, the mixtures of schemes 11 and 4 (Supplementary Table S3) were selected for foundation and backfill materials, respectively. It is pertinent to note that the slurry, which will be filled into the open pit, should also be mixed well and stirred homogeneously. Then it is poured into the open pit with a height of 5 cm, which corresponds to the foundation height of 10 m. Two iron plates of suitable sizes were fixed on the model at the same height as the OPB operation by appropriate clips to prevent the slurry from flowing out (Figure 4e). When the slurry gradually hardened, the initial strain value was adjusted consistent with the strain value at the end of the pouring to maintain the continuity of the strain curve. In order to observe strain trends of the OPB process in different stages clearly and facilitate the management and implementation of OPB, backfill heights were divided into 100, 40, and 40 m levels to reach the maximum OPB height of 190 m. In this way, the corresponding heights of 53, 21, and 21 cm can be calculated. This issue is presented in Figure 5, where stages 0 to 4 reflect the total OPB heights of 10, 110, 150, and 190 m, respectively.

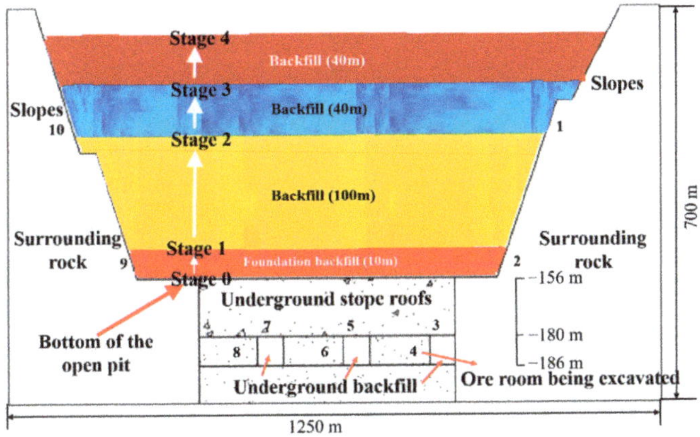

Figure 5. Schematic diagram of backfilling level.

In all experiments, a resistance strain soil pressure cell (BYD-1 type, Sanda Test Instrument, China) was applied to gather the strain data during OPB with a sampling frequency of 2 Hz (In the Section 3, the displacement is used to characterize the change in strain). Sampling was conducted until the maximum OPB height was finished. These steps were repeated constantly to achieve OPB sequentially to 58, 79 and 100 cm, and the iron plates was placed in the same way and raised constantly to the corresponding heights, which were consistent with the height of the OPB (Figure 4f). Finally, the model was cured under the same conditions for 28 days, when the mixtures hardened clearly.

3. Results and Discussion

3.1. The Influence of OPB on the Stope Roofs

Supplementary Table S4 presents the results of the numerical simulation. According to the safety factors of the roofs obtained from Equations (1) and (2), Table 5 shows the analysis of the variance (ANOVA) of the orthogonal test results [27]. In these analyses, F_1 can be coded as the Fisher value, which is called the F-value of the roofs obtained by the numerical simulation. The meanings of sum of square, degree of freedom, and mean square can be obtained in the related literature [28].

It should be indicated that the ANOVA selects 0.05 as the significance level (e.g., the reliability is more than 95%) to meet the analysis requirements. Searching in the Fisher distribution (referred to as F-distribution from statistical distributions according to GB 4086.4—83) shows that an $F_{0.05}(4, 8)$ value of 3.84 can be obtained.

Table 5. The ANOVA of the safety factor of the roofs obtained by the numerical simulation.

Source of Variance	Sum of Square	Degree of Freedom	Mean Square	F_1
Foundation strength (A) (MPa)	3.5259	4	0.88149	0.94
Foundation height (B) (m)	4.1103	4	1.02759	1.1
Backfill strength (C) (MPa)	3.0371	4	0.75929	0.81
Backfill height (D) (m)	23.7268	4	5.93171	6.35

Significance D > B > A > C.

3.1.1. The Stress Change

Based on the results of ANOVA provided in Table 5, it is observed that only the F-value of the backfill height (i.e., 6.35) is greater than 3.84. Therefore, the backfill height from 100 to 180 m significantly affects the safety factor of the roofs, while others have a smaller impact than that of backfill height on roofs. The significance weight ranking of four factors is: D > B > C > A (Table 5), which demonstrates that special attention should be paid to backfill height in the process of engineering practice to prevent negative effects on safety. Simultaneously, the trend of safety factors obtained is shown in Figure 6.

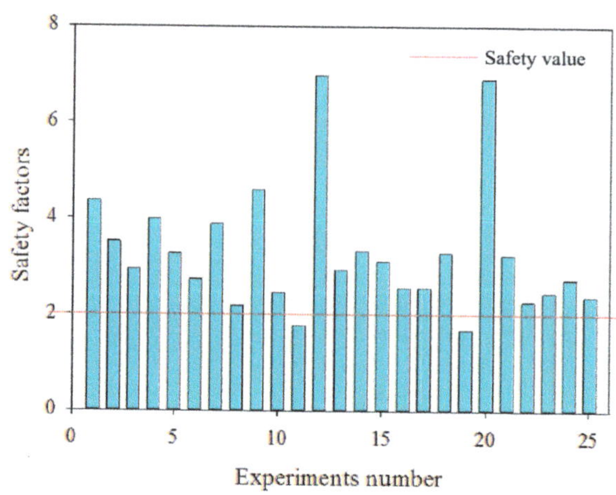

Figure 6. Safety factors of the roofs obtained by numerical simulation.

3.1.2. The Displacement Change

Figure 7 shows the overall displacement distributions of the model established in the horizontal and vertical directions. Figure 7a,b show that subsidence is inevitably caused on the roofs with the excavation of underground ore bodies, which is consistent with the results obtained by previous studies [29,30]. Moreover, it is found that the maximum downward displacement, which is 6.18 cm, is obtained at the center of the roofs corresponding to the center of stopes, which gradually decreases towards both sides.

3.1.3. Displacement Change in the Similar Model

Figure 8a,b show that from stage 0 to 2, the vertical displacement at point 5 increases as the height of OPB increases. It is found that the growth rate is initially higher, while it maintains a lower rate from stages 3 to 4. Therefore, the maximum displacement of 23.6 cm is obtained and the displacement value gradually becomes stable. Considering the horizontal displacement, it is found that from stages 0 to 2, the displacement produced by point 5 is consistent with that of the vertical displacement. It is worth noting that the maximum displacement of 14.2 cm is obtained during stages 3 to 4. Furthermore, no damage such as cracks and collapse, are found during the test. Therefore, the application of the proposed OPB cannot affect the safety of the roofs.

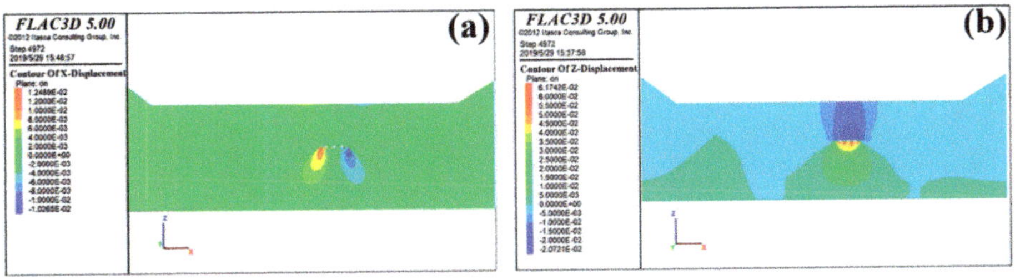

Figure 7. The overall displacements distributions in the horizontal (**a**) and vertical (**b**) directions.

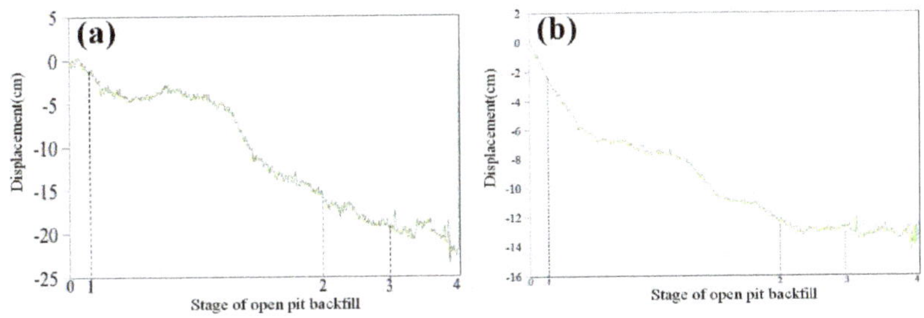

Figure 8. Displacements of measuring point 5 in the vertical (**a**) and horizontal (**b**) directions.

In Figure 6 and Supplementary Table S4, it should be noted that the values of safety factors in scheme 11 and 19 are lower than safety value, which is defined as the optimum factor of safety (The safety value is 2) according to safety requirements, demonstrating that these two options should not be allowed in situ application. However, obtained results show that OPB in underground mining does not threaten roofs, although certain stress and displacements are produced during this process [31].

3.2. The Influence of OPB on the Underground Backfill Pillars

3.2.1. The Stress Change

According to the safety factors of underground backfill pillars in Supplementary Table S4 and Figure 9, the results of ANOVA are summarized in Table 6. Where F_2 represents the F-value of the underground backfill pillars obtained by the numerical simulation with the same analysis conditions as roofs.

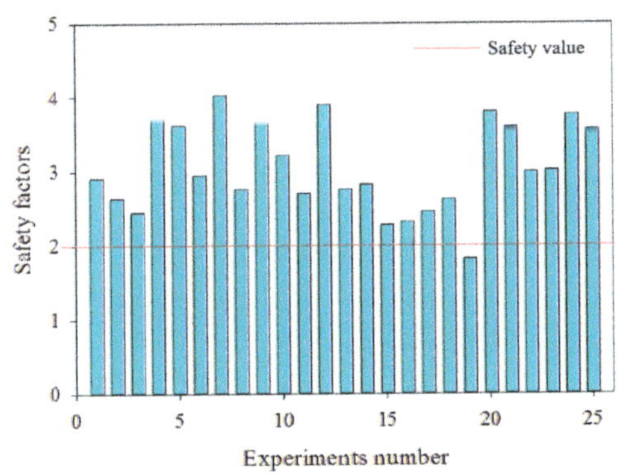

Figure 9. Safety factors of the underground backfill pillars obtained by numerical simulation.

Table 6. The ANOVA of the safety factor of the underground backfill pillars obtained by the numerical simulation.

Source of Variance	Sum of Square	Degree of Freedom	Mean Square	Value of F_1
Foundation strength (A) (MPa)	2.06478	4	0.51620	2.68
Foundation height (B) (m)	1.13730	4	0.28433	1.48
Backfill strength (C) (MPa)	2.17322	4	0.54331	2.82
Backfill height (D) (m)	1.63822	4	0.40956	2.13

Significance C > A > D > B.

Table 6 presents a fact that the F-values of all factors are less than 3.84. Therefore, the abovementioned factors slightly affect the safety and stability of the underground backfill pillars. This can be explained by the fact that a certain isolation effect is caused by roofs as an isolation layer. Therefore, the disturbance is weakened to underground backfill pillars [32].

3.2.2. The displacement Change

Figure 7 illustrates that larger displacements (6.1742 cm in the vertical direction) of the underground backfill pillars are produced in the vertical direction, which can be explained by the fact that underground backfill pillars have smaller stiffness than that of the surrounding ore bodies [33]. However, the overall structure is stable.

3.2.3. Displacement Change in Similar Model

Figure 10a,b show that as OPB progresses during stage 0 to stage 2, the pressure brought by increasing CPB gradually increases, leading to increased vertical displacement. Moreover, it is found that the displacement slightly fluctuates in the vertical direction when the OPB progresses from stages 2 to 3, and then it gradually rises. During stages 3 to 4, it levels off, with a maximum displacement value of 3.8 cm. It should be indicated that the horizontal displacement is mainly concentrated in stages 0 to 2. However, significant fluctuation can be found, resulting in the increased overall displacement. When entering stage 2, the horizontal displacement increases gradually at a declined rate. When entering stage 3, it remains stable with the maximum displacement value of 3.88 cm.

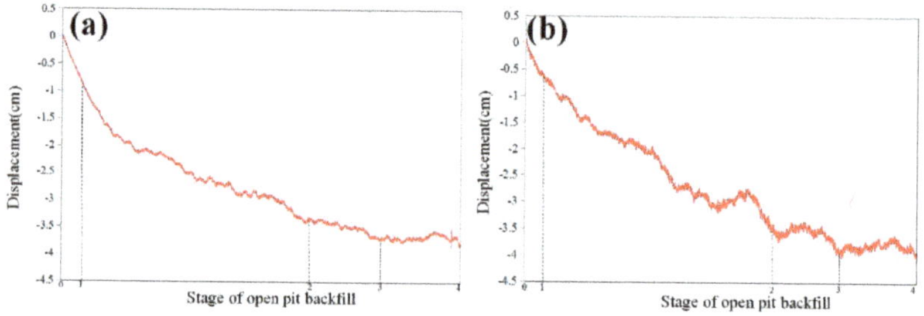

Figure 10. Displacements of measuring point 6 in the vertical (**a**) and horizontal (**b**) directions.

Based on the analysis of underground backfill pillars, it is found that the implementation of OPB will not exert a destructive effect on underground backfill pillars under conditions, which are consistent with roofs, despite the larger vertical displacements. However, the perfect integrity of a similar physical model can be observed. As revealed by Figure 9 and Supplementary Table S4, it is worth noting that the value of safety factor in scheme 19 is less than the safety value, which should be paid attention to in engineering operations.

3.3. The Influence of OPB on the Open-Pit Slopes

3.3.1. The Stress Change

Figure 11 presents the maximum and minimum principal stress distributions of the slopes of the open pit in the present study. In order to facilitate modeling, the inclination of the slope in Figure 11 does not seem to be as steep as that in Figures 2 and 5. Since it does not affect the final analysis, this difference can be ignored. It is observed that stress is mainly concentrated at the slope foot. However, the stress (1 MPa) can be found at the top of the slope, which conforms to the general principle of slope deformation [11]. As the height of OPB continuously increases, the stress concentration becomes more apparent and is accompanied by the expansion of the radiation range. Moreover, it is found that compressive stress is distributed in layers in the direction of gravity, which is consistent with the general discipline of compressive stress field distribution described by relevant literature [34]. Despite the concentration of the compressive stress in the slope foot, the overall structure can remain stable and does not cause hazards to the sloping rock, which is consistent with the previous study [35].

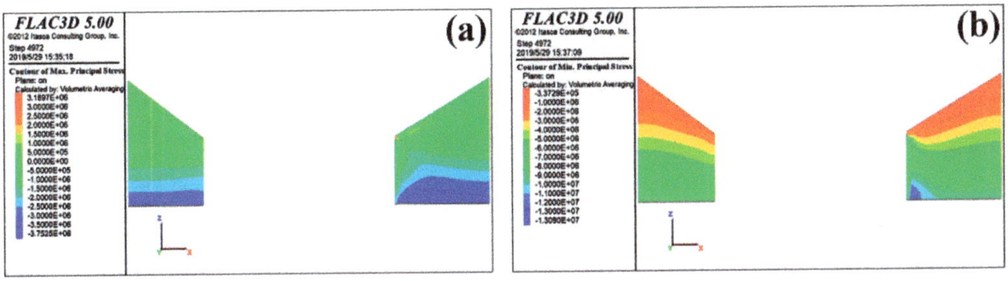

Figure 11. Slope: maximum (**a**) and minimum (**b**) principal stress distributions.

3.3.2. The Displacement Change

Figure 7 shows that no obvious displacement is produced on the slopes of both horizontal and vertical directions. This is the result of the reinforcement for slope foot given by overlying CPB, enhancing the anti-sliding force of the slopes. Therefore, the structure is more stable. The mechanism of stability and safety is discussed in the related literature [36].

3.3.3. Displacement Change in the Similar Model

Figure 12a shows that during stages 0 to 2, the vertical displacement gradually increases, and then becomes stable accompanied by sharp increments in displacement, which can reach up to 19 cm. This phenomenon is explained by errors during data measurement. The displacement fluctuates slightly and subsequently levels off. Therefore, when the height of OPB reaches a certain value, it acts as a support for the slopes on both sides which is, in turn, responsible for the gradual disturbance reduction on the slopes. Figure 12b shows that with the continuous filling of the open pit, the horizontal displacement of point 1 is sensitive to the change of the backfill height to a great extent, resulting in a continuous increase in the displacement. Moreover, it is found that the horizontal displacement fluctuates more than the vertical displacement of point 1. This phenomenon accounts for the fact that the slope angles on both sides affect part of the stress provided by the overlying CPB.

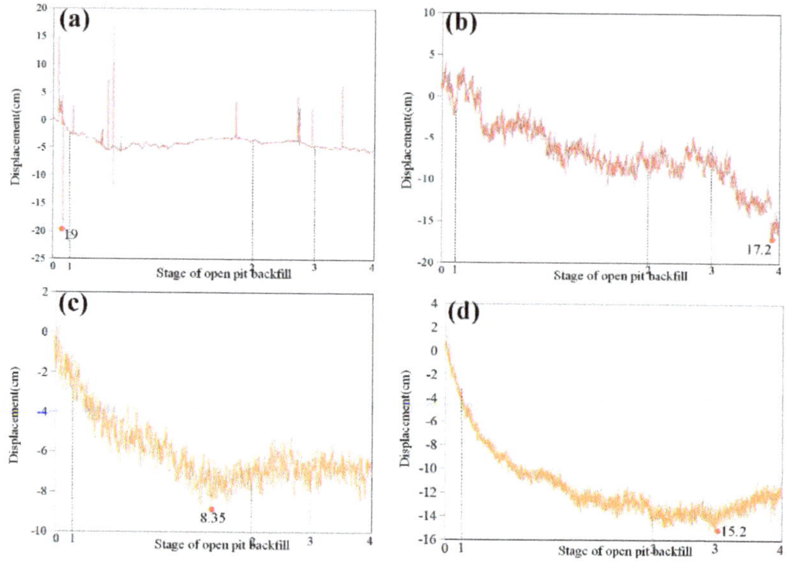

Figure 12. Displacements of measuring point 1 (**a**) vertical, (**b**) horizontal, and 2 (**c**) vertical, (**d**) horizontal.

Figure 12c shows the vertical displacement of point 2 increases as the height of OPB increases, obtaining a maximum displacement value of 8.35 cm. Then, it decreases slightly after the low point during stages 1 to 2, subsequently fluctuating steadily. In Figure 12d, the horizontal displacement of point 2 demonstrates a similar trend to its vertical displacement as revealed by Figure 12c. However, it is observed that the curve was smoother than that in Figure 12c. Therefore, there is a more stable rate of displacement change. Furthermore, at the beginning of stage 3, the maximum displacement value is obtained, which can reach 15.2 cm.

It is worth noting that the stress and displacements of the slopes can be restricted within a certain range, despite the presence of stress concentration. Therefore, the safety and stability of the slopes are verified.

According to the analysis of numerical simulation on the roofs, underground backfill pillars and slopes, the safety of the abovementioned three regions can be ensured significantly. Considering the use of OPB in many metal mines and relevant costs [37–42], selected parameters of OPB are foundation strength, foundation height, backfill strength and backfill height, which are 4 MPa, 10 m, 1.5 MPa, and 120 m, respectively. Moreover, corresponding safety factors obtained from the roofs and underground backfill pillars are 2.74 and 3.77, respectively, which meet the safety requirements. A similar physical model is investigated to verify whether selected parameters of OPB are reliable based on the numerical simulation. The results confirm that the disturbance effects on the three regions mainly occur before the OPB height is between 110 and 150 m. It should be indicated that no damage, such as cracks or collapse, was found during the test. Therefore, the application of the proposed method in the open pit does not affect the safety of the overall structure.

Comparing numerical simulations with similar physical model tests, it can be concluded that the results of the two are generally similar. It should be noted that similar physical model experiments can verify the accuracy of the numerical simulation results more intuitively.

4. Conclusions

In the present study, a technology named CPB is proposed, which has been used to fill open pits. Numerical simulation is performed to analyze the stress and displacement in the three mentioned areas. Moreover, the safety and stability during the process of

OPB are verified by physical model tests. The following conclusions are drawn from the present study:

(1) Comprehensive analysis is obtained through ANOVA. Among the factors affecting the roofs and underground backfill pillars, the backfill height plays a significant role in roofs and underground backfill pillars. Moreover, the obtained safety factors meet the requirements. In other words, the safety of the roofs and underground backfill pillars during OPB is guaranteed. It should be indicated that the foundation strength, foundation height, backfill strength, and backfill height are 4 MPa, 10 m, 1.5 MPa, and 120 m, respectively.

(2) The displacement gradually stabilizes when the backfill height is greater than 150 m. Moreover, no cracks or other damage appears, which verifies the reliability and accuracy of obtained numerical simulation results attained. In other words, the safety of the main regions, including the roofs, underground backfill pillars, and the slopes can be guaranteed during OPB.

The proposed method is utilized in waste tailings disposal and the OPB in a safe way, which can provide valuable suggestions for open-pit metal mines with similar situations. Simultaneously, the findings of this study provided a practical basis for the OPB in the future, which can treat solid waste resources to a large extent and realize safe production. Tailings, ordinary Portland cement 425#, and tap water, as aggregates, cementitious materials and mixing agent, respectively, are recommended for practical use of CPB. In the actual mixed design of CPB, it is necessary for engineers to detect the chemical composition of tailings to prevent danger to the environment and humans.

However, the results of the present study are restricted to the selection of the proportion of similar materials for the OPB and proportions of CPB components in practical application have not been determined so as to obtain optimized strengths, which can be further studied in future research.

Supplementary Materials: The following are available online at https://www.mdpi.com/article/10.3390/min11080818/s1, Table S1: Backfill parameters of the open pit implemented previously; Table S2: Uniform design ratios; Table S3: Results of uniform parameter tests; Table S4: Simulation results of safety factors of roofs and underground backfill pillars.

Author Contributions: Writing—original draft preparation, Q.Z. and B.Z.; data curation, B.Z. and D.W.; conceptualization, Q.Z. and B.Z.; methodology, Q.C., D.W. and X.G.; software, Q.Z., X.G. and Q.C.; validation, B.Z. and Q.C.; writing—review and editing, Q.Z. and B.Z.; supervision, Q.C.; funding acquisition, Q.Z. All authors have read and agreed to the published version of the manuscript.

Funding: This research was funded by Hunan Provincial Natural Science Foundation Project (No. 2020JJ5718), and the National Science Foundations of China (No. 52074351 and No. 52004330).

Acknowledgments: The authors would like to thank Hengliang Liu, Shihua Dong, and Weihong Su for funding support in the project, and Yikai Liu for technical support in the writing process.

Conflicts of Interest: The authors declare no conflict of interest.

References

1. King, B.; Goycoolea, M.; Newman, A. Optimizing the open pit-to-underground mining transition. *Eur. J. Oper. Res.* **2017**, *257*, 297–309. [CrossRef]
2. Dos Santos, T.B.; Lana, M.S.; Pereira, T.M.; Canbulat, I. Quantitative hazard assessment system (Has-Q) for open pit mine slopes. *Int. J. Min. Sci. Technol.* **2019**, *29*, 419–427. [CrossRef]
3. Qi, C.; Fourie, A. Cemented paste backfill for mineral tailings management: Review and future perspectives. *Miner. Eng.* **2019**, *144*, 106025. [CrossRef]
4. Wang, Q.; Li, J.; Zhu, X.; Yao, G.; Wu, P.; Wang, Z.; Lyu, X.; Hu, S.; Qiu, J.; Chen, P.; et al. Approach to the management of gold ore tailings via its application in cement production. *J. Clean. Prod.* **2020**, *269*, 122303. [CrossRef]
5. Sun, W.; Wang, H.; Hou, K. Control of waste rock-tailings paste backfill for active mining subsidence areas. *J. Clean. Prod.* **2018**, *171*, 567–579. [CrossRef]
6. Rimélé, M.A.; Dimitrakopoulos, R.; Gamache, M. A stochastic optimization method with in-pit waste and tailings disposal for open pit life-of-mine production planning. *Resour. Policy* **2018**, *57*, 112–121. [CrossRef]

7. Dong, K.; Xie, F.; Wang, W.; Chang, Y.; Lu, D.; Gu, X.; Chen, C. The detoxification and utilization of cyanide tailings: A critical review. *J. Clean. Prod.* **2021**, *302*, 126946. [CrossRef]
8. Chakilam, S.; Cui, L. Effect of polypropylene fiber content and fiber length on the saturated hydraulic conductivity of hydrating cemented paste backfill. *Constr. Build. Mater.* **2020**, *262*, 120854. [CrossRef]
9. Qi, C.; Fourie, A.; Chen, Q.; Zhang, Q. A strength prediction model using artificial intelligence for recycling waste tailings as cemented paste backfill. *J. Clean. Prod.* **2018**, *183*, 566–578. [CrossRef]
10. Benzaazoua, M.; Bussière, B.; Demers, I.; Aubertin, M.; Fried, É.; Blier, A. Integrated mine tailings management by combining environmental desulphurization and cemented paste backfill: Application to mine Doyon, Quebec, Canada. *Min. Eng.* **2008**, *21*, 330–340. [CrossRef]
11. Bastola, S.; Cai, M.; Damjanac, B. Slope stability assessment of an open pit using lattice-spring-based synthetic rock mass (LS-SRM) modeling approach. *J. Rock Mech. Geotech. Eng.* **2020**, *12*. [CrossRef]
12. Paul, S. Finite Element Analysis in Fused Deposition Modeling Research: A Literature Review. *Measurement* **2021**, *178*, 109320. [CrossRef]
13. Tu, H.; Zhou, H.; Lu, J.; Gao, Y.; Shi, L. Elastoplastic coupling analysis of high-strength concrete based on tests and the Mohr-Coulomb criterion. *Constr. Build. Mater.* **2020**, *255*, 119375. [CrossRef]
14. Zhang, J.; Wang, Z.; Song, Z. Numerical study on movement of dynamic strata in combined open-pit and under-ground mining based on similar material simulation experiment. *Arab. J. Geosci.* **2020**, *13*, 785. [CrossRef]
15. Wang, J.; Ren, L.; Xie, L.; Xie, H.; Ai, T. Maximum mean principal stress criterion for three-dimensional brittle fracture. *Int. J. Solids Struct.* **2016**, *102*, 142–154. [CrossRef]
16. Liu, B.; Zhou, J.; Wen, X.; Guo, J.; Deng, Z.; Hu, X. Mechanical performance and failure criterion of coral concrete under combined compression-shear stresses. *Constr. Build. Mater.* **2021**, *288*, 123050. [CrossRef]
17. Ma, D.; Zhang, J.; Duan, H.; Huang, Y.; Li, M.; Sun, Q.; Zhou, N. Reutilization of gangue wastes in underground backfilling mining: Overburden aquifer protection. *Chemosphere* **2021**, *264*, 128400. [CrossRef]
18. Khaboushan, A.S.; Osanloo, M.; Esfahanipour, A. Optimization of open pit to underground transition depth: An idea for reducing waste rock contamination while maximizing economic benefits. *J. Clean. Prod.* **2020**, *277*, 123530. [CrossRef]
19. Tian, Q.; Zhang, J.; Zhang, Y. Similar simulation experiment of expressway tunnel in karst area. *Constr. Build. Mater.* **2018**, *176*, 1–13. [CrossRef]
20. Tao, Z.; Shu, Y.; Yang, X.; Peng, Y.; Chen, Q.; Zhang, H. Physical model test study on shear strength characteristics of slope sliding surface in Nanfen open-pit mine. *Int. J. Min. Sci. Technol.* **2020**, *30*, 421–429. [CrossRef]
21. Li, Y.; She, L.; Wen, L.; Zhang, Q. Sensitivity analysis of drilling parameters in rock rotary drilling process based on orthogonal test method. *Eng. Geol.* **2020**, *270*, 105576. [CrossRef]
22. Morris, S.; Jackson, I. Implications of the similarity principle relating creep and attenuation in finely grained solids. *Mater. Sci. Eng. A* **2009**, *521*, 124–127. [CrossRef]
23. Han, C.; Zhang, J. Study on well hard shut-in experiment based on similarity principle and erosion of ram rubber. *Eng. Fail. Anal.* **2013**, *32*, 202–208. [CrossRef]
24. Ning, J.-H.; Zhou, Y.-D.; Fang, K.-T. Discrepancy for uniform design of experiments with mixtures. *J. Stat. Plan. Inference* **2011**, *141*, 1487–1496. [CrossRef]
25. Feng, Y.; Yang, Q.; Chen, Q.; Kero, J.; Andersson, A.; Ahmed, H.; Engström, F.; Samuelsson, C. Characterization and evaluation of the pozzolanic activity of granulated copper slag modified with CaO. *J. Clean. Prod.* **2019**, *232*, 1112–1120. [CrossRef]
26. Cardoso, V.H.; Caldas, P.; Giraldi, M.T.R.; Frazão, O.; de Carvalho, C.J.R.; Costa, J.C.; Santos, J.L. Experimental investigation of a strain gauge sensor based on Fiber Bragg Grating for diameter measurement. *Opt. Fiber Technol.* **2021**, *61*, 102428. [CrossRef]
27. Naik, A.B.; Reddy, A.C. Optimization of tensile strength in TIG welding using the Taguchi method and analysis of variance (ANOVA). *Therm. Sci. Eng. Prog.* **2018**, *8*, 327–339. [CrossRef]
28. Sei, T.; Shibata, H.; Takemura, A.; Ohara, K.; Takayama, N. Properties and applications of Fisher distribution on the rotation group. *J. Multivar. Anal.* **2013**, *116*, 440–455. [CrossRef]
29. Huang, G.; Kulatilake, P.H.; Shreedharan, S.; Cai, S.; Song, H. 3-D discontinuum numerical modeling of subsidence incorporating ore extraction and backfilling operations in an underground iron mine in China. *Int. J. Min. Sci. Technol.* **2017**, *27*, 191–201. [CrossRef]
30. Zingano, A.; Weiss, A. Subsidence over room and pillar retreat mining in a low coal seam. *Int. J. Min. Sci. Technol.* **2019**, *29*, 51–57. [CrossRef]
31. Hai, P.; Jian, Z. Research on protecting the safety of buildings by using backfill mining with solid. *Procedia Environ. Sci.* **2012**, *12*, 191–198. [CrossRef]
32. Lu, Y.; Li, M.; Wang, D.; Shi, X.; Li, H.; Zhu, Y.; Ye, Q.; Lu, J. Discharge and ignition characteristics from indentation fracture of coal mine roof. *Fuel* **2021**, *291*, 120208. [CrossRef]
33. Asem, P.; Gardoni, P. Bayesian estimation of the normal and shear stiffness for rock sockets in weak sedimentary rocks. *Int. J. Rock Mech. Min. Sci.* **2019**, *124*, 104129. [CrossRef]
34. Su, Z.; Shao, L. A three-dimensional slope stability analysis method based on finite element method stress analysis. *Eng. Geol.* **2021**, *280*, 105910. [CrossRef]

35. Kinakin, D.; Stead, D. Analysis of the distributions of stress in natural ridge forms: Implications for the deformation mechanisms of rock slopes and the formation of sackung. *Geomorphology* **2005**, *65*, 85–100. [CrossRef]
36. Wu, Z.; Chen, C.; Lu, X.; Pei, L.; Zhang, L. Discussion on the allowable safety factor of slope stability for high rockfill dams in China. *Eng. Geol.* **2020**, *272*, 105666. [CrossRef]
37. Zheng, J.; Li, L. Experimental study of the "short-term" pressures of uncemented paste backfill with different solid contents for barricade design. *J. Clean. Prod.* **2020**, *275*, 123068. [CrossRef]
38. Zhang, Q.; Li, Y.; Chen, Q.; Liu, Y.; Feng, Y.; Wang, D. Effects of temperatures and pH values on rheological properties of cemented paste backfill. *J. Cent. South Univ.* **2021**, *28*, 1707–1723. Available online: https://kns.cnki.net/kcms/detail/43.1516.tb.20210318.1429.002.html (accessed on 27 July 2021). [CrossRef]
39. Xu, W.-B.; Liu, B.; Wu, W.-L. Strength and deformation behaviors of cemented tailings backfill under triaxial compression. *J. Central South Univ.* **2020**, *27*, 3531–3543. [CrossRef]
40. Guo, Y.-X.; Ran, H.-Y.; Feng, G.-R.; Du, X.-J.; Qi, T.-Y.; Wang, Z.-H. Effects of curing under step-by-step load on mechanical and deformation properties of cemented gangue backfill column. *J. Central South Univ.* **2020**, *27*, 3417–3435. [CrossRef]
41. Chen, Q.; Sun, S.; Liu, Y.; Qi, C.; Zhou, H.; Zhang, Q. Experimental and numerical study on immobilization and leaching characteristics of fluoride from phos-phogypsum based cemented paste backfill. *Int. J. Min. Metall. Mater.* **2021**. [CrossRef]
42. Li, X.; Peng, K.; Peng, J.; Xu, H. Effect of Cyclic Wetting–Drying Treatment on Strength and Failure Behavior of Two Quartz-Rich Sandstones under Direct Shear. *Rock Mech. Rock Eng.* **2021**. [CrossRef]

minerals

MDPI

Article

Experimental Study on Mechanical Properties of Paste Backfill with Flue-Gas Desulphurisation Gypsum under Combined Action of Dry–Wet Cycles and Chloride Erosion

Sheng Wang, Feng Wang *, Dawei Yin, Tianqi Jiang and Zhen Zhang

State Key Laboratory of Mine Disaster Prevention and Control, Shandong University of Science and Technology, Qingdao 266590, China; 202082010032@sdust.edu.cn (S.W.); yindawei@sdust.edu.cn (D.Y.); Jiang_Tianqi@hotmail.com (T.J.); 201882010040@sdust.edu.cn (Z.Z.)
* Correspondence: wangfeng@sdust.edu.cn; Tel.: +86-137-8985-9661

Abstract: Flue-gas desulphurisation gypsum—a solid waste from power plants—can be used to prepare paste backfill for reducing costs. Most paste backfills are exposed to dry–wet cycles and chloride salt-rich water in mines. Therefore, the mechanical properties and damage mechanisms of paste backfill with desulphurised gypsum under the coupling action of erosion due to chloride with different concentrations and dry–wet cycles were investigated using methods such as visual observation, mass measurement, uniaxial compression, acoustic emission, Fourier-transform infrared spectroscopy, X-ray diffraction analysis, and field-emission scanning electron microscopy. With an increasing number of dry–wet cycles, the mass, elastic modulus, and strength of the paste backfill exhibited the trend of increasing first and then decreasing. The failure mechanism changed from mainly vertical fractures to the alternating development of vertical and horizontal fractures. The surface denudation effect of the specimens in a solution with a higher concentration was more severe under the same number of dry–wet cycles. In this study, the laws governing the mass change, strength change, degree of surface denudation, and failure pattern of desulphurised gypsum-filled specimens under different concentrations of chloride salt and different numbers of dry–wet cycles were derived.

Keywords: backfilling material; desulphurised gypsum; chloride erosion; dry–wet cycles; mechanical properties

Citation: Wang, S.; Wang, F.; Yin, D.; Jiang, T.; Zhang, Z. Experimental Study on Mechanical Properties of Paste Backfill with Flue-Gas Desulphurisation Gypsum under Combined Action of Dry–Wet Cycles and Chloride Erosion. *Minerals* **2021**, *11*, 882. https://doi.org/10.3390/min11080882

Academic Editor: Abbas Taheri

Received: 14 July 2021
Accepted: 13 August 2021
Published: 15 August 2021

1. Introduction

Mining with backfilling has been adopted by many mines because of the ability to control mine subsidence and reduce ecological damage [1,2], but the cost of filling restricts its widespread application. Paste backfill is widely used in coal mines as one of the main materials for filling voids [3]. However, problems such as the gradual reduction in filling materials and high prices have severely limited the application of paste filling in coal mines. There is an urgent need to find a filling material that is inexpensive, widely available, and satisfies all requirements.

Filling materials have been investigated by scholars worldwide. Feng et al. [4] examined the effect of construction waste used as a coarse aggregate in filling paste. Chen et al. [5] studied various mechanical properties of red mud after it was added to filling paste using macroscopic and mesoscopic test methods, and the results provided a basis for the reasonable use of red mud. Li et al. [6] added kaolin to filling paste and studied its mechanical properties and failure characteristics. Cheng et al. [7] studied the effect of coarse aggregate gradation on the yield stress of the filling paste, proposed a paste stability coefficient to characterise the paste gradation, and found that a slurry with a large paste stability coefficient has a compact and stable honeycomb structure that allows the paste to overcome the relatively high yield stress. Fall et al. [8] investigated the strength and microscopic changes of the filling body with Portland cement as a cementing agent under the coupled effects of sulphate and temperature. Liu et al. [9] revealed the influence of the gangue particle size on the compressive strength of specimens and summarised the mechanism of solidification of heavy metals inside the raw materials through gangue-based cementitious filling materials. Zhang et al. [10] studied the mesoscopic structural failure characteristics of a crushed gangue filling material and revealed the impact mechanism of the compression deformation of granular filling materials under loading.

The aforementioned studies provided a variety of filling materials for coal mining. Gypsum can be added to the filling paste system to stimulate the activity of fly ash. Additionally, coal-fired power plants produce a large amount of desulphurised gypsum in the process of flue-gas desulphurisation. After simple dichlorination, the desulphurised gypsum can be added to the filling paste instead of natural gypsum [11], which reduces the cost of the filling paste while disposing the industrial waste generated by coal-fired power plants. Chandar [12] and Hou et al. [13] experimentally verified the possibility of replacing natural gypsum with desulphurised gypsum. Tzouvala et al. [14] demonstrated that desulphurised gypsum can prolong the setting time and increase the strength of cement by comparing desulphurised gypsum with natural gypsum experimentally. Colak et al. [15] studied a composite gel material prepared with cement and desulphurised gypsum and found that the desulphurised gypsum significantly increased the strength of the material. Guo et al. [16,17] confirmed the reuse value of desulphurised gypsum by studying the effects of desulphurised gypsum as an admixture on the mechanical properties and durability of cement. Zhang et al. [18] studied a waterproof material prepared with fly ash, blast furnace slag, and desulphurised gypsum and proved that desulphurised gypsum could improve the water resistance of the paste. Qiao et al. [19] combined cement, desulphurised gypsum, and fly ash at a certain ratio into a composite gel material with excellent cementitious properties. The studies performed by scholars worldwide prove that filling paste mixed with desulphurised gypsum fulfils all the requirements of mining with filling.

A large amount of mine water exists in the goaf, and there are several ions dissolved in the mine water, among which Cl^- is abundant [20]. Mine water significantly affects the performance of the filling paste. Zhang et al. [21] found that chloride has a positive effect on the early strength of cement paste. In another study, Zhang et al. [22] synthesised AFt (ettringite) and, using XRD (X-ray diffraction) analysis, thermal analysis, and FTIR (Fourier-transform infrared) spectroscopy, determined that the binding effect of AFt with Cl^- involved physical adsorption and chemical binding. Ekolu [23] believed that, at high Cl^- concentrations, AFt can chemically bond with Cl^- to produce Friedel's salt. Du et al. [24] studied the erosion and deterioration effects of chloride solutions with different mass fractions on filling paste with different ages. Qiao et al. [25] studied the effects of four cations, i.e., Na^+, K^+, Mg^{2+}, and Ca^{2+}, on the binding mechanism between Cl^- and concrete and found that the binding capacity exhibited the following trend: $Ca^{2+} > Mg^{2+} > Na^+ \approx K^+$. Farnam et al. [26] studied the effect of Cl^- on the transport properties of concrete and found that the Friedel's salt generated by the reaction between the two sealed the concrete pores, degrading the transport properties. Chen et al. [27] studied the effects of different initial Cl^- concentrations on the early strength of gangue-filled paste and found that an initial Cl^- concentration of 10% resulted in the highest early strength. Wang et al. [28] found that, when Cl^- is introduced into cementitious materials, the binding of cementitious materials with Cl^- can be divided into chemical binding and physical adsorption, where the chemical binding mainly consists of the formation of Friedel's salt.

The rise and fall of the groundwater level expose the filling paste to a dry–wet cycle environment [29], reducing the strength of the filling paste. Li et al. [30] studied the erosion patterns of Cl^- on cementitious composite materials under dry–wet cycles and found that dry–wet cycles increased the erosion rate of Cl^-. Chen et al. [31] studied the erosion effects of chloride and sulphate on concrete under the action of dry–wet cycles and found that the presence of SO_4^{2-} in the chloride solution inhibited the formation of Friedel's salt, and that the presence of Cl^- in the sulphate solution inhibited the formation of sulphate products such as AFt. Aldaood et al. [32] studied the deterioration of concrete by dry–wet cycles with different gypsum contents and found that the deterioration was most pronounced at a gypsum content of 25%. However, no studies have been performed on the mechanical properties and deterioration of paste backfill with desulphurised gypsum under the coupled effects of dry–wet cycles [33] and chloride erosion [34].

Because the application of new materials must ensure that the filling body can maintain a certain strength under the special environment of chloride salt erosion and dry–wet cycles, a filling with a uniaxial compressive strength of >2 MPa is typically considered to satisfy the requirements of filling mining. Thus, paste backfill with desulphurised gypsum was prepared in this study. Methods such as appearance analysis, mass determination, uniaxial compression, AE (acoustic emission), and other mesoscopic analysis techniques, including FTIR spectroscopy, XRD analysis, and FE-SEM (field-emission scanning electron microscopy), were used to analyse the changes in the paste backfill with desulphurised gypsum in terms of the appearance, mass, uniaxial compression, elastic modulus, and fracture development of desulphurisation under dry–wet cycles in solutions of different concentrations. The results revealed the mechanical properties and damage mechanisms of paste backfill with desulphurised gypsum under the coupled effects of erosion due to chloride at different concentrations and dry–wet cycles. The study verifies whether the strength of the filling-paste sample in this special environment can still meet the strength required by filling mining. As the lack of filling materials is a global problem, this paper presents the changes in the mechanical properties of desulphurised gypsum as a filling material in the special environment of underground mining, providing theoretical support for the application of desulphurised gypsum in paste filling. Therefore, this research has practical significance and reference value.

2. Test Materials and Methodology

2.1. Specimen Preparation

Blast furnace slag is usually added as an admixture to help gypsum increase the strength of the paste backfill [35], and its optimum ratio was selected for the preparation of paste backfill with desulphurised gypsum in accordance with a related study performed by Jiang [36]. The paste backfills were mainly prepared using cement, fly ash, coal gangue, blast furnace slag, desulphurised gypsum, and water. Grade 32.5 ordinary silicate cement, produced by Shandong Rizhao No. 3 Cement Plant, was used in the test. Fly ash and desulphurised gypsum were taken from the Zoucheng Power Plant of Huadian Power International in the Shandong Province. The coal gangue was taken from the Daizhuang coal mine after secondary crushing. The granulated blast furnace slag used in the test was obtained from Qingdao Special Steel Co. The composition of the raw materials was determined via X-ray fluorescence spectroscopy (Panalytical Axios FAST, Malvern, UK), and the main chemical compositions of the raw materials are listed in Table 1. The morphologies of the raw materials are shown in Figure 1.

The fraction of raw materials used in the test was cement/gangue/admixture = 1:6:4, where the admixture ratio was fly ash/blast furnace slag/gypsum = 12:3:1. Thirty-nine paste backfills with desulphurised gypsum were prepared with this ratio, of which three specimens (the control group) were not subjected to dry–wet cycles.

Table 1. Raw material chemical compositions for the backfill sample.

Raw Materials	Chemical Composition and Content (%)					
	SiO_2	Al_2O_3	CaO	MgO	SO_3	Fe_2O_3
Cement	22.30	4.82	62.47	2.34	1.82	3.83
Fly ash	56.30	18.05	11.84	3.01	1.19	4.93
Coal gangue	59.83	20.10	4.89	1.22	-	6.52
Blast furnace slag	35.83	12.73	38.09	7.86	-	2.79
Desulphurised gypsum	24.40	2.62	32.25	1.33	37.58	0.73

Cement | Fly ash | Waste rock (in coal mining) | High furnace slag | Desulphurised gypsum

Figure 1. Photographs of raw materials.

The mass fraction for all the above solids was 78%, and the maximum particle size of the coal gangue was 25 mm. The preparation process was as follows: first, the raw materials were weighed using an electronic scale and placed in NJ-160 Agitator in accordance with the above ratio for stirring and mixing. After complete mixing, water was added and continued stirring until it became a paste filling body with certain fluidity Then, the prepared filling paste was poured into a 70.7 mm × 70.7 mm × 70.7 mm cube mould. By manual vibration and tamping, the bubble entrainment of the sample in the mould was removed. After 24 h, the solidified specimens were demoulded and taken out. The prepared specimens were placed in a box at a constant temperature and humidity (temperature of 20 °C and 95% relative humidity) to be maintained for 28 days [37]. Subsequently, the mechanical properties of the specimens were tested using GB/T50107-2010.

2.2. Test Methodology

Chloride solutions with mass fractions of 3%, 5%, and 10% were prepared. After 28 days of conditioning, the specimens were placed in containers with each of the three solutions. The dry–wet cycles were simulated by immersing the specimen for 12 h, followed by 12 h of drying in an oven at 60 °C [38]. The specimens were divided into four groups and subjected to five, 10, 15, and 20 dry–wet cycles. The number of dry–wet cycles was labelled as 1, 2, 3, and 4, and the concentration of the solutions was labelled as 3, 5, and 10. For example, the specimen subjected to three dry–wet cycles in a solution with a 3% mass fraction was labelled as C13. This nomenclature was used for all the specimens. The masses of the specimens before and after the test were weighed and recorded using an electronic scale with a precision of 0.01 g. The mechanical properties of the specimens at the end of the dry–wet cycles were tested using a Shimadzu AG-X250 (Shimadzu, Kyoto, Japan) electronic universal tester with a loading rate of 0.01 mm/s. The uniaxial compression failure process was monitored in real time using an AE test system (MISTRAS PCI-2 Series, Physical Acoustics, Princeton, NJ, USA). Figure 2 shows the procedure for the mechanical test. The main amplification was set as 40 dB, the threshold was 45 dB, the floating threshold was 6 dB, the probe harmonic frequency was 100–600 kHz, and the sampling frequency was 106 times/s. Vaseline was applied between the sensor and the specimen for coupling to reduce the acoustic impedance difference at the contact surface and the reflection-induced loss of energy at this interface to ensure that the AE signal was well received by the sensor. A FTIR instrument (Nicolet iS5, Thermo Fisher Scientific, Waltham, MA, USA) was used to identify the surface functional groups of the specimen samples, and a total of 50 scans were taken to obtain the infrared spectra. Crystal calibration was performed on the specimen powders using a X-ray diffractometer (Bruker D8 advance, Bruker, Billerica, MA, USA) with a diffraction-angle range of 5–90° and a scan step of 8 °/min. The mesostructures of the bulk specimens of backfill paste were scanned using FE-SEM (Apreo S Hivac, Thermo Fisher Scientific, Waltham, MA, USA) with resolutions of 5000×, 10,000×, and 40,000×. The samples were sprayed with gold before the scanning.

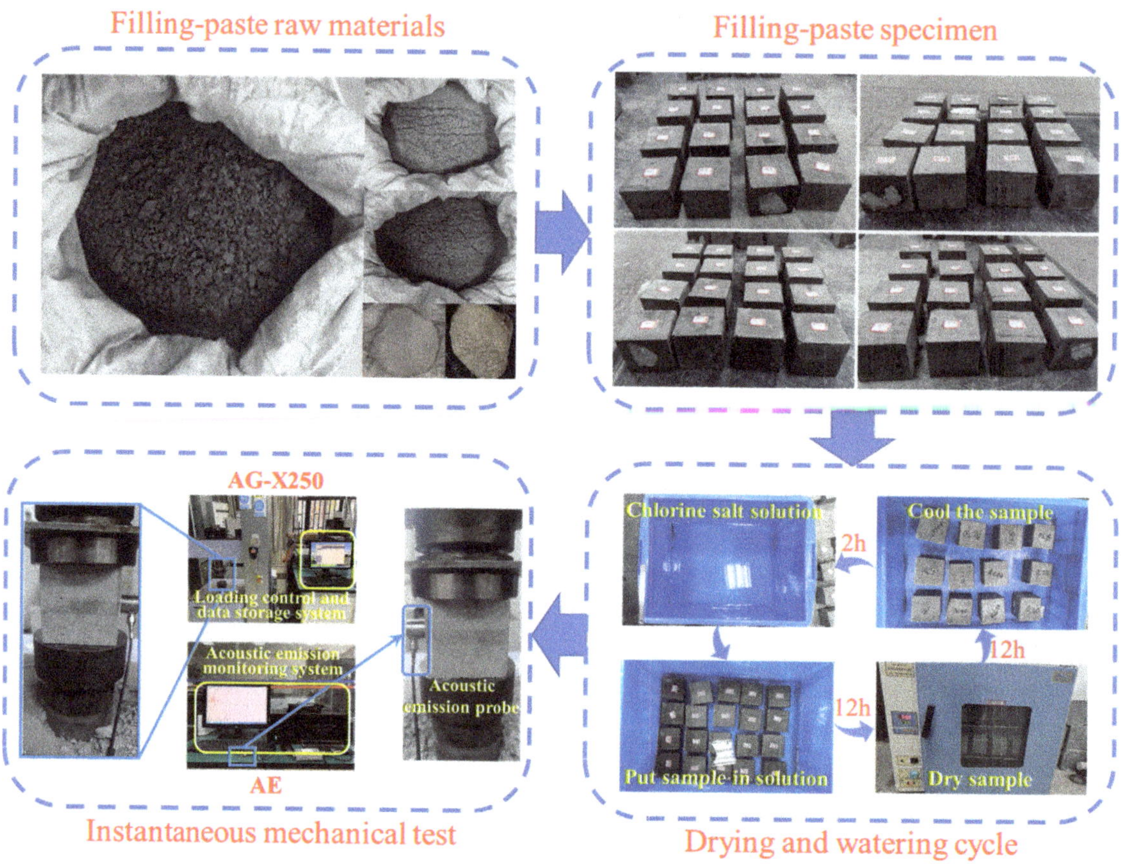

Figure 2. Mechanical performance test flow.

3. Mechanical Performance Tests

3.1. Effects of Dry–Wet Cycles Coupled with Chloride on Masses of Filled Specimens

The most apparent effects of dry–wet cycles on the paste backfills with desulphurised gypsum were the effects on their appearance and mass. The changes of the surfaces of the specimens in 3%, 5%, and 10% chloride solutions for five, 10, 15, and 20 dry–wet cycles are shown in Figure 3, and the mass changes are shown in Figure 4.

Gangue was used as a coarse aggregate in the preparation of the filling paste. The surface denudation effect under the coupled effects of dry–wet cycles and chloride erosion of the specimen led to the appearance of gangue on the specimen surface (observed as black spots). The area of gangue on the specimen surface can be used to evaluate the degree of deterioration of the specimen. As shown in Figure 3, gangue only appeared after 20 dry–wet cycles in the solution with a 3% concentration; gangue appeared after 15 and 20 dry–wet cycles in the solution with a 5% concentration, whereas the appearance of gangue started from the 10th dry–wet cycle in the solution with a 10% concentration. The surface denudation of the specimen was particularly exacerbated with the increasing number of dry–wet cycles. The area of black gangue was increasingly large at the edges and corners of the specimens, where passivation gradually occurred under the denudation effect. With an increase in the solution concentration, the surface denudation of the specimen was exacerbated, and the number of revealed gangue particles on the specimen surface gradually increased. This indicates that both the concentration of the chloride solution and the number of dry–wet cycles were positively correlated with the degree of surface denudation of the specimens. In addition to the deterioration of the specimen surface due to the dry–wet cycles, the Cl^- in the solution reacted with the dense AFt on the specimen surface to form insoluble Friedel's salt, which exacerbated the denudation of the specimen surface.

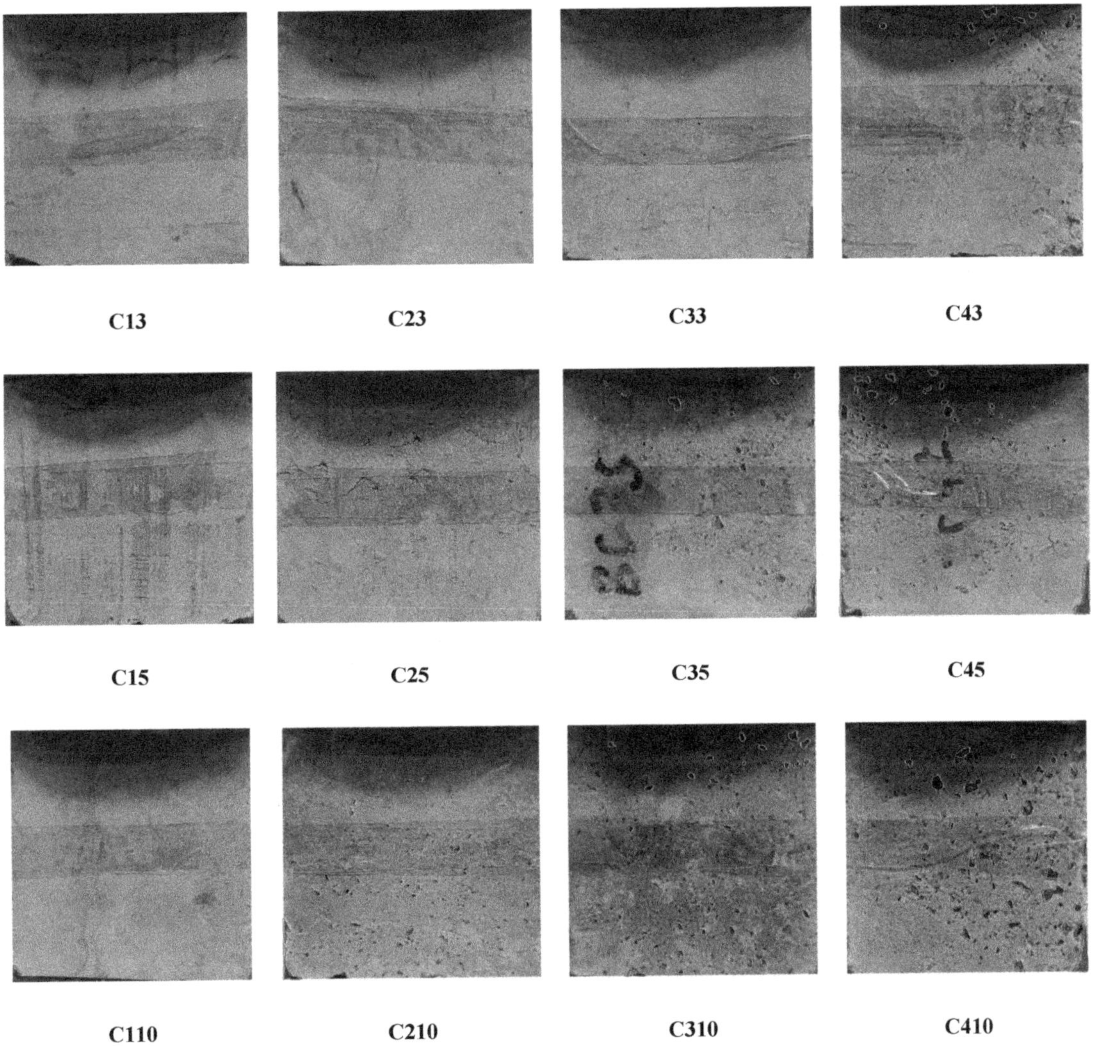

Figure 3. Deterioration of the surfaces of paste backfill with desulphurised gypsum under different conditions.

Figure 4 presents the relationship between the mass of the backfill sample and the number of dry–wet cycles at different concentrations. Here, the mass of the backfill without dry–wet cycles was labelled as 0, and the mass variation of the backfill after dry–wet cycles was plotted as a broken line chart. As shown, the mass changes for the filled specimens in 3%, 5%, and 10% chloride salt solutions all exhibited a pattern of decreasing, increasing, and then decreasing again. At the fifth and 10th dry–wet cycles, the masses of the specimens exhibited substantial reductions; the specimen in the 3% solution exhibited the largest mass loss (13.5 g). The components of the specimens that were not involved in the hydration reaction dissolved into the solution, while the loosely bound water in the sample evaporated in the drying phase of the dry–wet cycles, leading to a rapid decline in the masses of the specimens at the beginning of the dry–wet cycles. As the number of cycles increased, the masses of the specimens also began to increase, and, by the 15th dry–wet cycle, the masses of the specimens reached their maximum value. The specimen in the 10% solution exhibited the largest mass increase; it was 1.5 g heavier than the specimens not subjected to dry–wet cycles. During this process, the hydration inside the specimen proceeded further, and the free water in the solution solidified into bounded water inside the specimen, while the chlorine ions in the chloride salt solution and the specimens underwent a new hydration reaction. This new reaction filled the voids inside the specimens and improved the compactness of the specimens, increasing the masses of the specimens. lastly, the masses of the specimens began to decrease when the hydration reaction inside the specimens approached completion, and Cl^- reacted with the hydration products in the specimens to form Friedel's salt. The overall structure of Friedel's salt was looser than that of the hydration products; thus, the combined effects of the dry–wet cycles and chloride erosion led to the deterioration of the overall structures of the specimens. This resulted in the denudation of the surface layer and the loosening of the internal structures of the specimens, causing a reduction in mass.

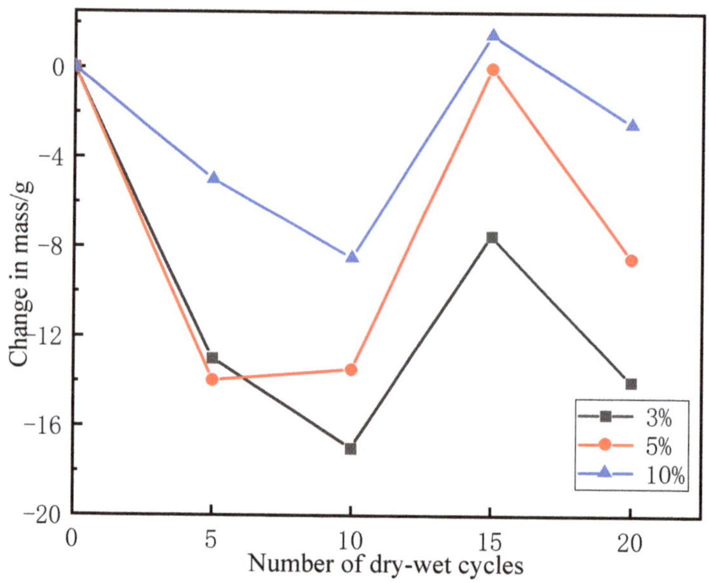

Figure 4. Variations of the backfill mass with respect to the number of dry–wet cycles at different solution concentrations.

Under the dry–wet cycles, the 3% chlorine salt solution had the most severe deterioration effect on the specimens (with the largest mass loss), followed by the 5% and 10% solutions. This is mainly because Cl^- at a high concentration can penetrate the specimen rapidly and activate fly ash inside the specimen [21], which accelerates the hydration reaction inside the specimen. Furthermore, the high concentration of Cl^- can maintain the Friedel's salt in the specimen and the chloride salt precipitated through the pores inside the specimen during the dry–wet cycles. This reduced the deterioration effect of the dry–wet cycles on the masses of the specimens. The continuation of hydration reaction inside the specimen is the main factor for the mass growth of the specimen. In the specimen with low concentration solution, the velocity of Cl^- entering the specimen was low, and the degree of hydration reaction inside the specimen was low, leading to serious mass loss of the specimen. When the number of drying and wetting cycle was 15 or 20, the hydration reaction inside the specimen was complete, and Cl^- reacted with hydration products to generate Friedel's salt. However, the low concentration of solution concentration was not enough to maintain the existence of Friedel's salt, which led to the decomposition of Friedel's salt inside the specimen and exacerbated the mass loss of the specimen.

3.2. Effects of Dry–Wet Cycles Coupled with Chloride Salt on Uniaxial Compressive Strengths of Backfill Specimens

Table 2 is Uniaxial strengths of desulphurisation gypsum-filled specimens. Table 3 is Elastic moduli of backfill specimens with desulphurised gypsum. Figure 5 presents the variations in the peak stresses of backfill samples subjected to uniaxial compression under different conditions. As shown, the peak stresses of the specimens under the uniaxial compression conditions for the 3%, 5%, and 10% chloride salt solutions increased slightly and then decreased. At the fifth and 10th dry–wet cycles, there was a slight increase in the strength of the specimen (approximately 12%). This is because the hydration reaction within the specimen was not yet complete, and the strength increase provided time for the hydration reaction within the specimen to continue. The chloride solution also enhanced the strength of the specimen, because the hydration reaction within the specimen was not yet completed at this time. Moreover, Cl^- can fully interact with the fly ash through diffusion and chemically react with the potential active substances Al_2O_3 and Ca^{2+} inside the specimen to generate gel substances such as C–A–H ($xCaO{\cdot}Al_2O_3{\cdot}yH_2O$) [39]. The newly generated substances fill the internal pores of the specimen, increasing its strength. Thus, the specimens exhibited an increase in the peak stress at the beginning of the dry–wet cycles. As the number of dry–wet cycles increased, the peak stress of the specimens began to decrease, and, by the 20th dry–wet cycle, the strength of the specimens was reduced by approximately 15%. This is because the hydration reaction inside the specimens was complete, and the Cl^- reacted with the non-hydrated active materials and the hydration products inside the specimen to produce Friedel's salt [40], which is insoluble and tends to swell. These hydration products accumulated and swelled inside the specimens, generating cracks in the specimens. Owing to the physical denudation of the specimens due to the dry–wet cycles, the AFt films generated on the outer surface of the specimens were damaged and peeled off [41]. Thus, Cl^- could more easily invade the specimen interior and precipitate in the voids, causing fracture enlargement, which further reduced the specimen strength. Therefore, the peak stresses of the specimens exhibited an overall reduction.

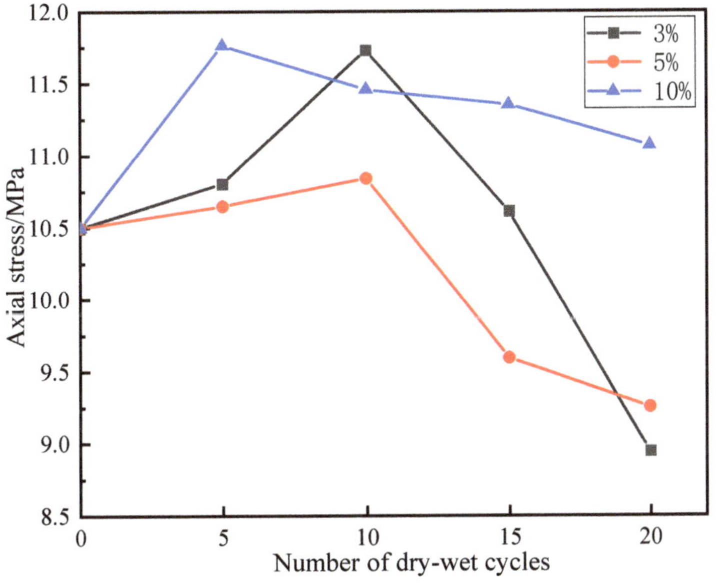

Figure 5. Variations in the peak stresses of backfill specimens under uniaxial compression.

Table 2. Uniaxial strengths of desulphurisation gypsum-filled specimens.

Solution Concentration / Number of Dry–Wet Cycles	Uniaxial Compressive Strength (MPa)											
	3%			Average	5%			Average	10%			Average
0	10.77	10.74	10.14	10.49	-	-	-	-	-	-	-	-
5	9.41	10.97	12.02	10.80	10.47	10.61	10.86	10.64	10.20	12.19	12.90	11.76
10	12.72	11.81	10.65	11.73	10.81	11.73	9.97	10.84	10.96	13.15	10.26	11.46
15	11.20	11.04	9.59	10.61	9.75	10.37	8.65	9.59	10.13	11.31	12.61	11.35
20	8.95	7.94	9.94	8.94	7.90	9.59	10.25	9.25	9.14	11.98	12.08	11.07

Table 3. Elastic moduli of backfill specimens with desulphurised gypsum.

Solution Concentration / Number of Dry–Wet Cycles	Elastic Modulus (MPa)											
	3%			Average	5%			Average	10%			Average
0	835.21	764.58	833.17	810.99	-	-	-	-	-	-	-	-
5	725.60	1019.87	902.74	882.74	945.18	819.49	912.33	892.33	756.21	1025.90	921.05	901.05
10	1044.36	931.05	1017.71	997.71	755.93	916.95	866.44	846.44	819.87	1030.47	955.17	935.17
15	942.89	886.97	944.93	924.93	884.44	888.26	916.35	896.35	627.94	1004.25	846.10	826.10
20	603.23	520.11	591.67	571.67	647.76	733.23	720.49	700.49	646.37	1096.05	901.21	881.21

The strengths of the specimens in the 10% chloride solution were maximised at the fifth dry–wet cycle (with a 12% increase in strength) and exhibited a decreasing trend as the dry–wet cycles continued. The change in the peak stress was the most moderate at this concentration, with a 5% increase in strength by the 20th cycle. This is because the higher concentration of Cl^- created a larger concentration gradient within the specimen interior, allowing Cl^- to enter the interior more quickly during the dry–wet cycles. The peak strength of the specimen was maximised at the fifth dry–wet cycle owing to the enhancement effect of Cl^-. The specimen strength slowly decreased in the subsequent dry–wet cycles because the high concentration of chloride salt maintained the presence of Friedel's salt [42,43], preventing the generation of a large number of fractures inside the specimen due to the decomposition of the Friedel's salt. Simultaneously, the components of the solution continued to precipitate within the pores inside the specimens, filling the fractures. This eventually reduced the specimen strength, although the reduction rate was lower. The specimens in the 3% and 5% chloride solutions reached their maximum peak strengths at the 10th dry–wet cycle. The peak stress then began to decrease, and the reduction rate was significantly higher for the 3% solution (from a 12% increase in strength at the 10th dry–wet cycle to a 15% reduction in strength at the 20th cycle) than for the other two concentrations. This was due to the limited chlorine salt concentration. Although Friedel's salt was available, the absence of high concentrations of Cl^- to sustain it inside the specimens resulted in numerous fractures inside the specimens after the decomposition of the Friedel's salt. The combination of this effect with the deterioration of the specimen strength due to the dry–wet cycles resulted in a sharp decrease in the specimen strength.

Figure 6 presents the changes in the elastic moduli of backfill samples under different conditions. As the number of dry–wet cycles increased, the elastic moduli of the specimens exhibited a trend of first increasing and then decreasing. At the fifth dry–wet cycle, the elastic moduli of the specimens exhibited an increase of approximately 10% for all the solution concentrations. At the 10th dry–wet cycle, the elastic moduli of the specimens continued to increase, with the specimens in the 3% solution exhibiting increases of up to 18%. The hydration reaction in the specimens proceeded further, closing the pores and fractures within the specimens to some extent and resulting in a more compact internal structure. This was eventually reflected by the increasing elastic moduli of the specimens. As the dry–wet cycles continued, the elastic moduli of the specimens all decreased to different degrees, with the specimens in the 3% solution exhibiting the largest reductions (30%). The elastic moduli of the specimens in the 5% solution decreased by 18%, and those of the specimens in the 10% solution decreased by only 3%. This is mainly because the hydration reaction inside the specimens was complete. Friedel's salt formed by the reaction between Cl^- and the hydration products led to the deterioration of the internal structure of the backfill. Furthermore, the low concentration of Cl^- could not maintain the Friedel's salt, leading to its decomposition. This caused deterioration of the internal structures of the specimens, which was reflected by reductions in the elastic moduli.

The uniaxial compressive strength and elastic modulus of the paste backfill with desulphurised gypsum exhibited an overall pattern of a short increase followed by a sharp decrease under the coupled effects of chloride erosion and dry–wet cycles. Despite the degradation of the specimens due to the chloride erosion and dry–wet cycles, the initial strength of the paste backfills with desulphurised gypsum was as high as 10.49 MPa, and the average strength of the group with the most severe deterioration after the dry–wet cycles reached 8.94 MPa. Thus, desulphurised gypsum can satisfy the strength requirements for application as a filler material in mining with paste backfill.

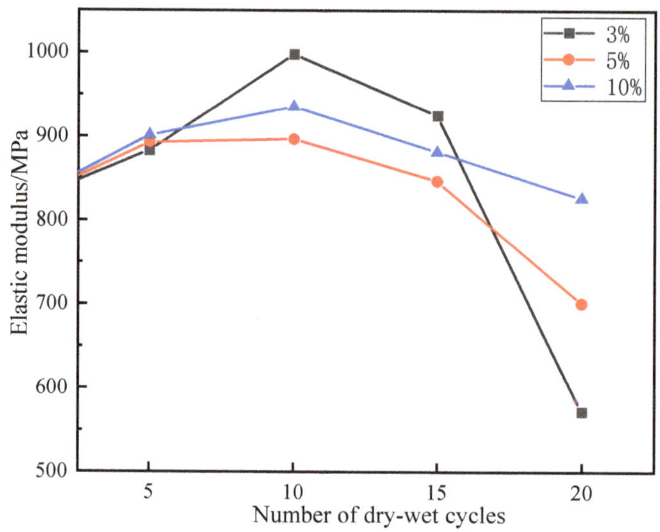

Figure 6. Variations in the elastic moduli of backfill specimens.

3.3. AE Characteristics

AE (acoustic emission) is often used as a non-destructive monitoring method to analyse the deformation of a specimen during loading or the generation of elastic acoustic waves during fracturing and to, thus, determine the mechanical characteristics of its internal mesostructure [44]. However, the influence of the variables of this test on the unstable failure processes of the specimens mainly affects the initiation, expansion, and penetration processes of microcracks in the specimens during loading by changing their mesostructures, compositions, and other characteristics. The specimens will inevitably have voids during the preparation process. Thus, fissures and weak cementitious surfaces will be produced around the gangue during the hardening process of the specimens, which leads to a variety of macroscopic failure forms, such as tensile failures and shear failures. The AE test results generally indicated a high frequency and low strength, and the fracture development of the specimens was dominated by vertical edge cracks.

Japanese scholars Shiotani et al. [45] calculated the RA (the ratio of rise time to amplitude) value of AE for bending and shear tests on rocks and concluded that low RA values correspond to shear fractures and high RA values correspond to tension fractures. In combination with the uniaxial compression process, AE can be similarly analysed according to the four stages of uniaxial compression. In the first PDS (pressure density stage), more AE signals were generated in the PDS, as the paste backfill differed from the rock specimens; thus, multiple shear failures occurred. Because no external load was applied during the paste preparation, the internal structures of the specimens were relatively loose compared with the rocks, having a large number of internal pores and fractures. Under loading, the internal pores closed, generating acoustic signals. In the subsequent LES (linear elastic deformation stage), the internal pores and fractures in the specimens were closed in the PDS, and the applied load was not sufficient to create new fractures in the specimen; thus, few AE signals were monitored at this stage, and the tension failure was dominant. The internal structures of the specimens were relatively stable. The AE signals had high intensities in the two subsequent stages; thus, shear failure dominated within the specimen. In the PPYS (pre-peak yielding stage), the specimen's structure underwent massive macroscopic failures as the load increased and new fractures began to form inside the specimen; thus, the AE signals monitored at this stage were strong. In the PPDS (post-peak damage stage), the damage to the overall structure of the filling reduced its load-bearing capacity. A more pronounced deformation also occurred at this stage, and numerous AE signals were generated, accompanied by the macroscopic deformation of the specimen.

Figure 7 presents the macroscopic fracture development and AE monitoring results for the specimens during uniaxial compression. As shown, the RA values measured via AE monitoring of the specimens during uniaxial compression at the fifth and 10th dry–wet cycles were low. The RA values increased suddenly when the specimens reached the peak stress, at which time the vertical fractures dominated the failure processes of the specimens. This indicates that, at the fifth and 10th dry–wet cycles, the specimens had good integrity in the early stage of uniaxial compression and did not have dense fractures. Numerous fractures were produced in the specimens when the peak stresses were reached. From a mesoscopic perspective, this is because the chloride salt enhanced the hydration reaction inside the specimens at the beginning of the dry–wet cycles. The precipitation of chloride salt in the pores inside the specimens led to relatively compact specimens, and the fractures were filled with the hydration products and chloride salt crystals, increasing the brittleness of the specimens. Therefore, few fractures were produced in the early stage of stress application. When the peak stresses of the specimens were reached, numerous fractures were instantaneously produced under the load. At the 15th and 20th dry–wet cycles, the RA values obtained via the AE monitoring of the specimens no longer exhibited large sudden changes but tended to gradually increase in the fracture extension stage, and the uniformity of the increase in the RA values was more obvious at the 20th dry–wet cycle. At this time, the macroscopic fracture development of the specimens mostly involved the combined development of vertical and horizontal fractures. From a mesoscopic perspective, this is because, with the increase in the number of dry–wet cycles, the internal deterioration of the specimens by the coupled effects of chloride salt and dry–wet cycles became increasingly severe. Numerous fractures were generated inside the specimens at this time, which gradually expanded and generated new fractures under the load. This led to an increase in the RA values over time in the stage before the peak stress was reached. Additionally, the internal loosening of the specimens contributed to the simultaneous development of vertical and horizontal fractures. More intense AE signals were detected for the specimens at the 15th and 20th dry–wet cycles than at the fifth and 10th dry–wet cycles, indicating that more fractures were developed in the specimens under the load during the early dry–wet cycles. This is because the original fractures closed by hydration within the specimens were reopened under the load, generating AE signals. The interiors of the specimens were further damaged as the dry–wet cycles progressed. Moreover, the Friedel's salt formed by the erosion of chloride salt had good swelling properties. Its large accumulation inside the specimens increased the plasticity of the specimens. The enhancement of the plasticity of the specimens reduced the AE signals generated by internal brittle fractures, which eventually led to more AE signals being detected at the fifth and 10th dry–wet cycles than at the 15th and 20th dry–wet cycles. This phenomenon was more prominent for solutions with higher concentrations, because they could maintain the presence of more Friedel's salt in the specimens, resulting in a higher plasticity.

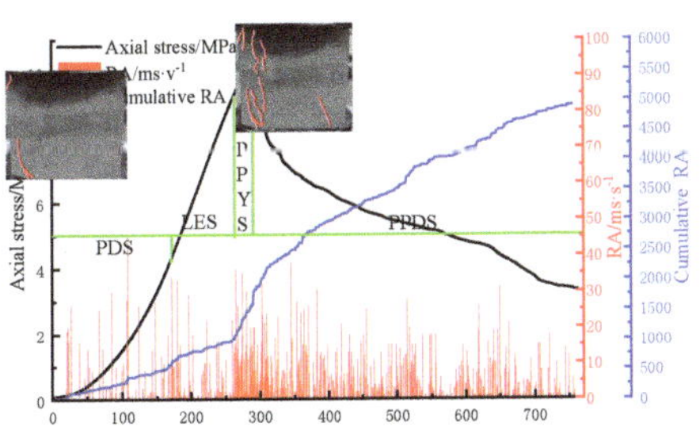

C13

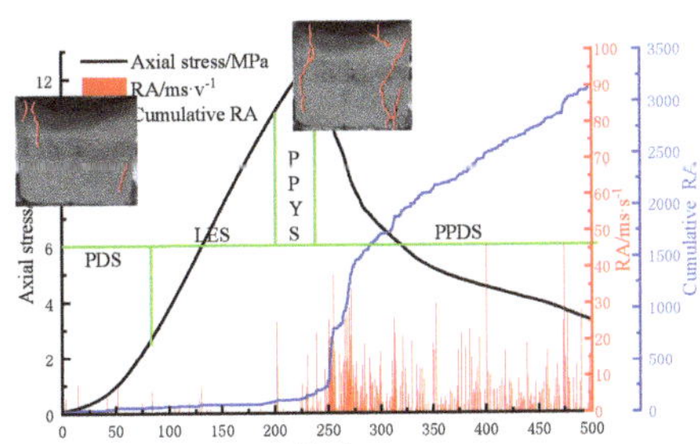

C23

Figure 7. *Cont.*

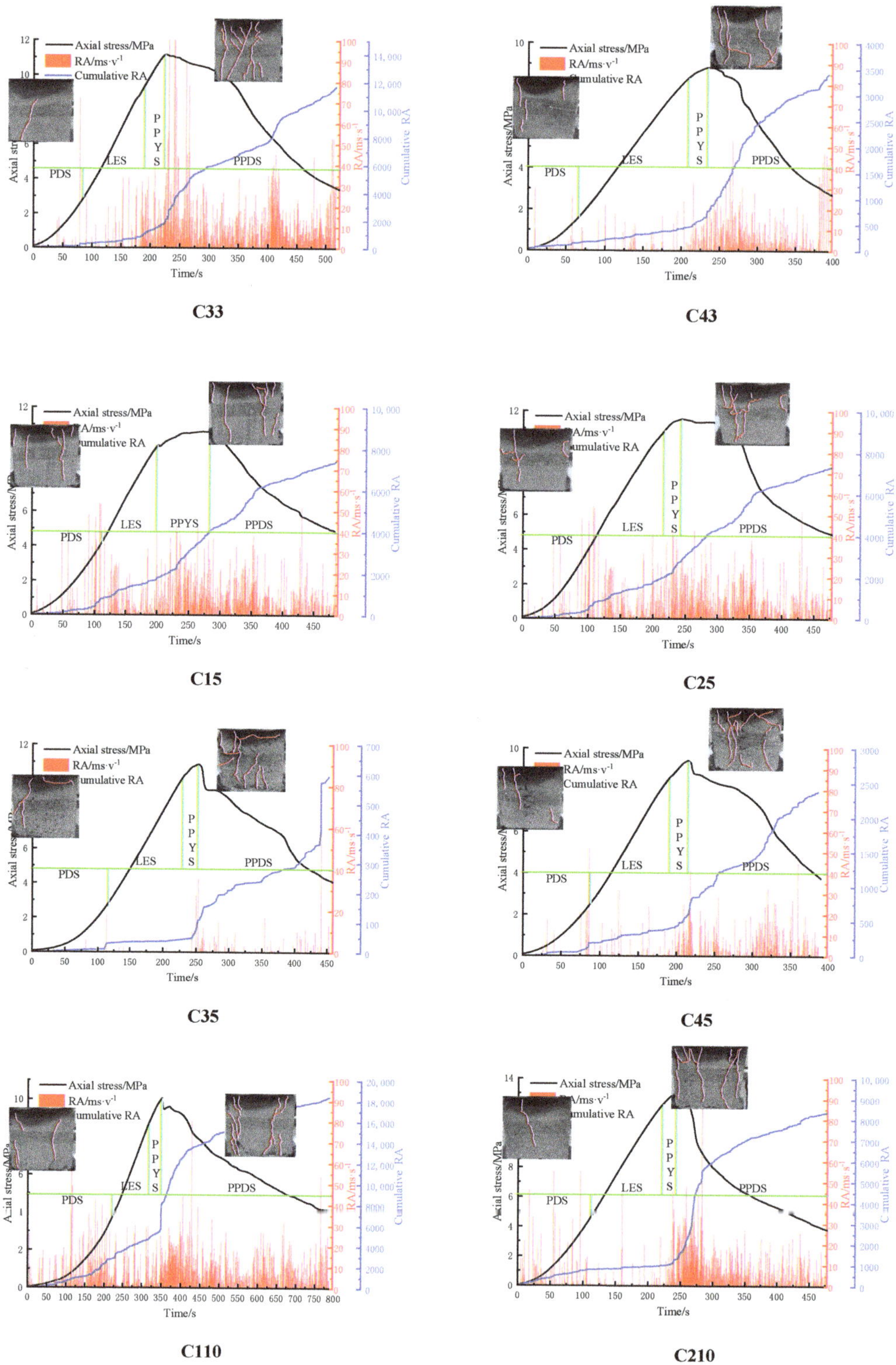

Figure 7. *Cont.*

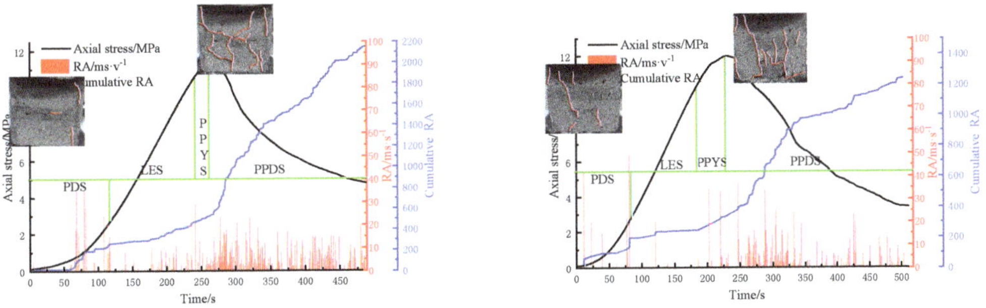

C310 **C410**

Figure 7. AE monitoring during uniaxial compression.

4. Mesoscopic Tests

The macroscopic mechanical property tests revealed that the effects of the dry–wet cycles with the 5% solution on the mass, mechanical properties, and AE of the paste backfill specimens were between those for the 3% and 10% solutions. Additionally, the dry–wet cycling with the 5% solution did not produce unique phenomena that were not observed for the other groups. Thus, the mesoscopic analysis was only conducted for the specimens in the 3% and 10% solutions. The test equipment is shown in Figure 8.

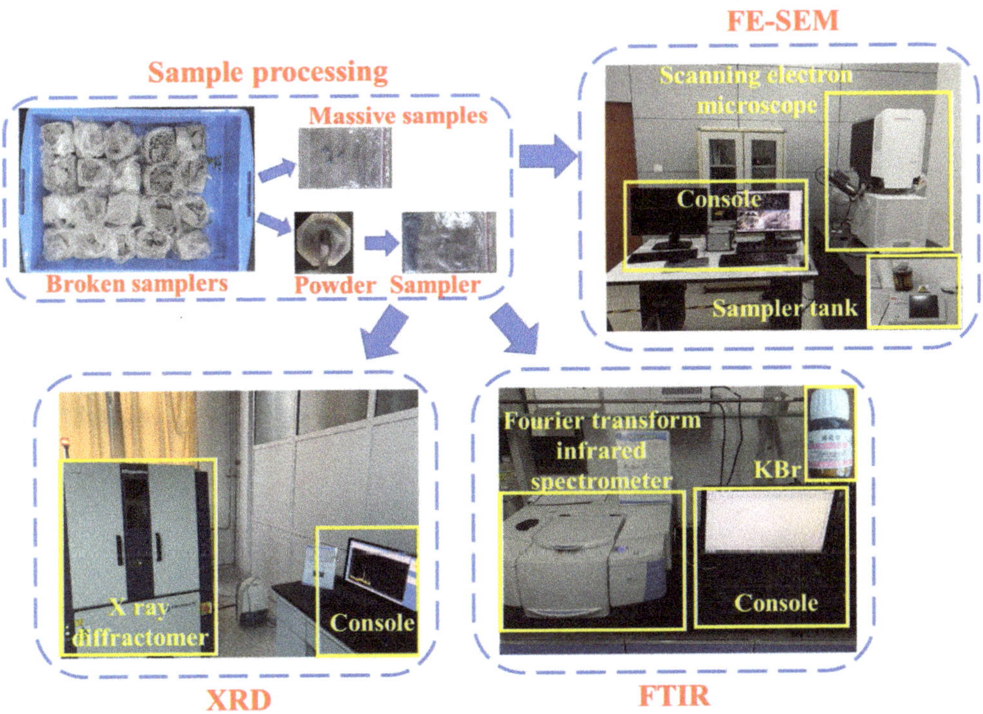

Figure 8. Mesoscopic test equipment.

4.1. Analysis of Surface Functional Groups via FTIR Spectroscopy

Figure 9 shows the FTIR spectra of the filled specimens. The absorption peak at 3630 cm^{-1} corresponded to the O–H stretching vibration of $Ca(OH)_2$. The absorption band at 1683–1648 cm^{-1} corresponded to the bending vibration of H–O. The absorption band at the 1058–1110 cm^{-1} corresponded to the asymmetric stretching vibration of S–O. As indicated by the XRD results, these absorption peaks were the vibration peaks of AFt. There was a difference at 780 cm^{-1}, which was attributed to the absorption peak of the reaction that produced Friedel's salt, according to the XRD analysis. The broad absorption spectrum band at 3303–3330 cm^{-1} corresponded to the stretching vibration of Si=OH in the C–S–H ($xCaO{\cdot}SiO_2{\cdot}yH_2O$) gel. The peak at 1448–1458 cm^{-1} may have been due to the stretching vibration of the C=C absorption band. The weak absorption peak at 875 cm^{-1} characterises the bending vibration of Si=OH in the C-S-H gel, and that at 471 cm^{-1} characterises the bending vibration of Si–O–Si in the C–S–H gel. As indicated by the XRD results, the main hydration products were $Ca(OH)_2$, C–S–H, and C–A–H gel, while $Ca(OH)_2$ underwent a chemical reaction with CO_2 in the air to produce $CaCO_3$ [27,46].

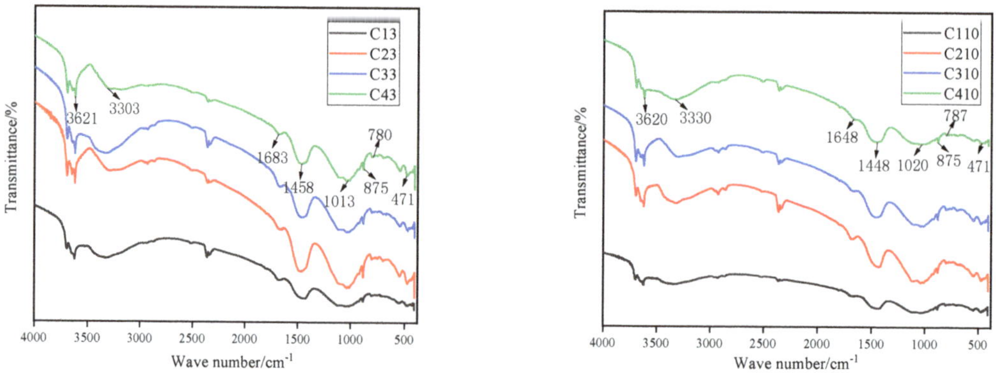

Figure 9. FTIR spectra of filled specimens.

4.2. Crystal Calibration Analysis with XRD Diffraction

Figure 10 shows the calibration of the spectra after the XRD test. The XRD diffraction spectra of the paste backfill specimens were calibrated for SiO_2, $Al_6Si_2O_{13}$, AFt, C_3SH ($3CaO{\cdot}SiO_2{\cdot}yH_2O$), C_4AH ($4CaO{\cdot}Al_2O_3{\cdot}H_2O$), $CaCO_3$, and Friedel's salt. As the fly ash, gangue, and slag in the specimen contained large amounts of SiO_2, the XRD diffraction spectra of the specimens were very prominent in the calibration of SiO_2. $Al_6Si_2O_{13}$ (mullite) was also present in the fly ash and gangue. The main hydration products of cement, fly ash, and other raw materials included C_3SH, C_4AH, and other substances. These hydration products were in a gel state; therefore, the coarse and fine aggregates in the specimens were cemented to each other and solidified the internal fractures in the specimens, which was one of the main reasons for the increased strengths of the specimens. AFt is generated by the chemical reaction among flue-gas desulphurisation gypsum, C_3A($3CaO{\cdot}Al_2O_3$), and C_4AF in cement. It is an extremely insoluble, highly crystalline mineral that can quickly consolidate enough free water in the cementing material and grow in the form of needles to form a dense and stable structure, significantly increasing the early strength of the specimen. The CaO in the raw material of the normal hydration reaction reacted chemically with other substances in the specimens to produce the gel product. Thus, the amount of $Ca(OH)_2$ was insignificant, and $Ca(OH)_2$ appeared only as an intermediate product. However, Cl^- as a stimulant activated the fly ash in the specimen material and reacted with it to produce Friedel's salt under the action of chloride erosion. $Ca(OH)_2$ was produced in large quantities as a by-product of this reaction. The chemical reactions that produced $Ca(OH)_2$ are shown in Figures 9 and 10. However, in the presence of air, $Ca(OH)_2$ quickly reacts with CO_2 in the air, forming $CaCO_3$. Therefore, a large amount of $CaCO_3$ was present in the XRD spectra. Friedel's salt can also be generated after the reaction of Cl^- with the gel product inside the specimens. When the number of dry–wet cycles reached 15 or 20, the number of AFt diffraction peaks in the XRD spectra of the paste backfill specimens started to decrease, and diffraction peaks of the Friedel's salt formed by the reaction started to appear.

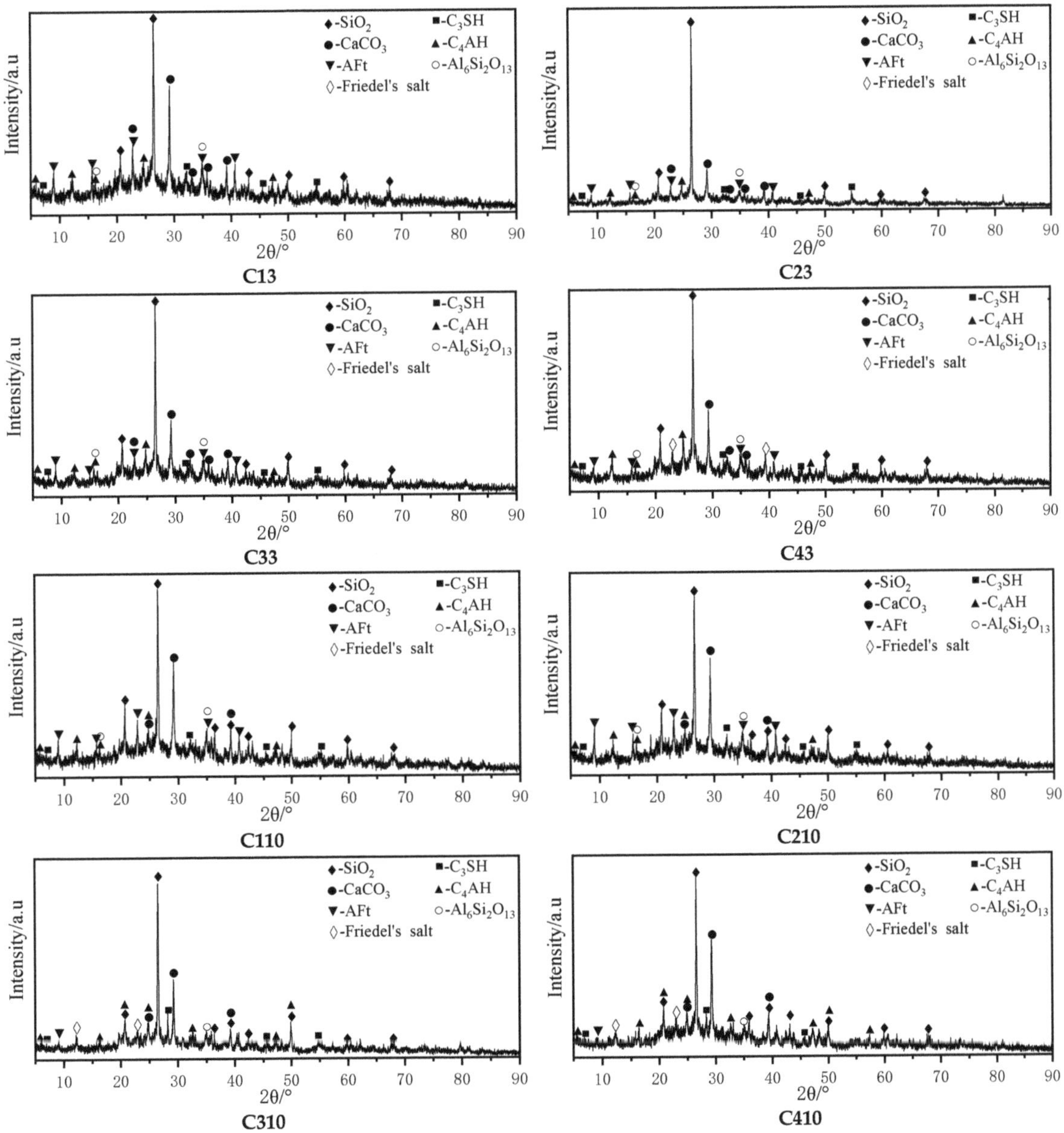

Figure 10. Sample XRD spectra.

4.3. FE-SEM Mesostructure Analysis

Figure 11 presents the surface morphologies of different test samples under a scanning electron microscope. As shown, the number of pores and fractures in the FE-SEM images of the specimens increased with the number of dry–wet cycles. The internal and surface structures became looser and more broken, respectively. Needle- and rod-like AFt was still observed in the specimens at the fifth and 10th dry–wet cycles. However, as the dry–wet cycles continued, the AFt started to transform into Friedel's salt in the presence of Cl^-. This is consistent with the phenomenon observed in the XRD spectrum calibration, where the AFt diffraction peak decreased and that of Friedel's salt started to appear at the 15th and 20th dry–wet cycles.

As shown in Figure 11, the number of pores and holes in the specimens in a higher-concentration solution was smaller than that for a lower-concentration solution at the same number of dry–wet cycles. Additionally, the surface structures of the specimens in the low-concentration solution were extremely fragmented and had an overall shattered shape, whereas the surfaces of the specimens in the high-concentration solution maintained their integrity relatively well. There was more AFt on the surfaces of the specimens in the higher-concentration solution, because the Cl^- reacted more completely with AFt to form Friedel's salt and better maintained the Friedel's salt so that it could exist stably in the interiors of the specimens.

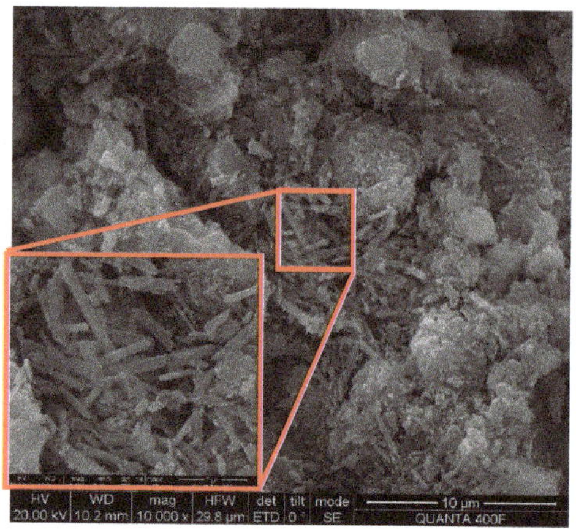

C13

C23

C33

C43

Figure 11. *Cont.*

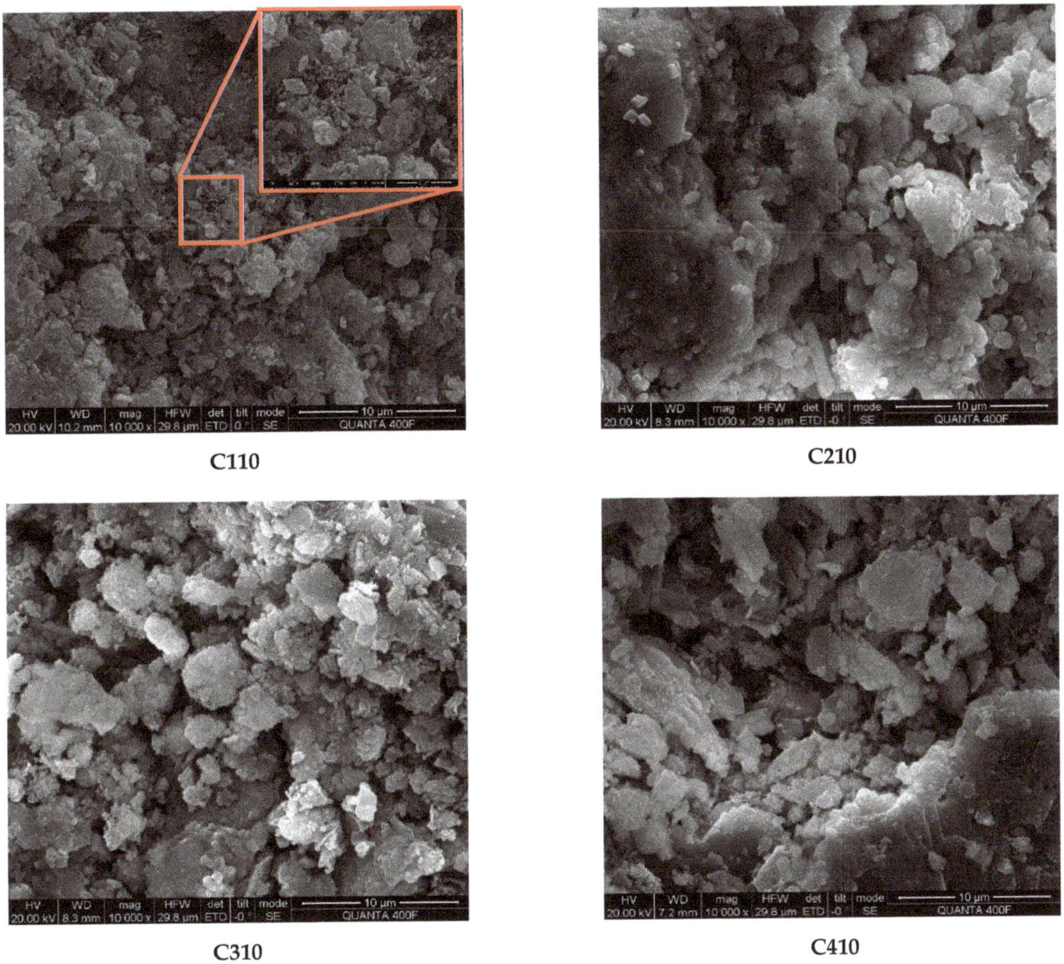

Figure 11. FE-SEM images of the filled specimens.

5. Discussion of Mechanisms

1. Chemical reactions of specimens after mixing

Among the materials of the paste backfill specimens, gangue and slag are incorporated into the filling body system as aggregates. Cement, as a binder, is the main component for ensuring the mechanical strength of the backfill specimens. Because fly ash is potentially active while acting as a fine aggregate, it can undergo a chemical reaction under the stimulus of gypsum to produce a hydraulic substance that enhances the mechanical strength of the paste backfill specimens. Equations (1)–(4) are the hydration reactions of each mineral of cement clinker with water; they occur after the contact of cement particles with water.

$$2(3CaO{\cdot}SiO_2) + 6H_2O \rightarrow 3CaO{\cdot}2SiO_2{\cdot}3H_2O + 3Ca(OH)_2 \quad (1)$$

$$2(3CaO{\cdot}SiO_2) + 4H_2O \rightarrow 3CaO{\cdot}2SiO_2{\cdot}3H_2O + Ca(OH)_2 \quad (2)$$

$$3CaO{\cdot}Al_2O_3 + 6H_2O \rightarrow 3CaO{\cdot}Al_2O_3{\cdot}6H_2O \quad (3)$$

$$4CaO{\cdot}Al_2O_3{\cdot}Fe_2O_3 + 7H_2O \rightarrow 3CaO{\cdot}Al_2O_3{\cdot}6H_2O + CaO{\cdot}Fe_2O_3{\cdot}H_2O \quad (4)$$

Gypsum is used as an additive to regulate the setting time and mechanical strength of cement. The main hydration products of the cementing system are AFt (ettringite), C–S–H gel, and C–A–H gel. Adding FCD gypsum, which has undergone heat treatment under certain conditions, can improve the physical properties of cement by improving its strength. Fly ash, such as volcanic fly ash and FCD gypsum, can significantly improve the compressive strength of a fly ash "slag" cement-based cementing system and increase the strength of cemented bodies [46]. Equations (5) and (6) are the relevant chemical reactions of fly ash under the action of FCD gypsum as a stimulant. Equation (7) represents the reaction of FCD gypsum in cement with C_3A to generate AFt, and Equation (8) represents the reaction of the gel material wrapped on the surface of fly ash particles with AlO^{2-}, Ca^{2+}, and $SO_4{}^{2-}$ present in the system to generate AFt.

$$Ca(OH)_2 + Al_2O_3 \rightarrow CaO{\cdot}Al_2O_3{\cdot}H_2O \quad (5)$$

$$Ca(OH)_2 + SiO_2 + H_2O \rightarrow CaO{\cdot}SiO_2{\cdot}2H_2O \quad (6)$$

$$3CaO{\cdot}Al_2O_3 + 3CaSO_4{\cdot}2H_2O + 26H_2O \rightarrow 3CaO{\cdot}Al_2O_3{\cdot}3CaSO_4{\cdot}32H_2O \quad (7)$$

$$AlO^{2-} + Ca^{2+} + OH^- + SO_4{}^{2-} \rightarrow 3CaO{\cdot}Al_2O_3{\cdot}3CaSO_4{\cdot}32H_2O \quad (8)$$

2. In the fifth and 10th dry–wet cycles, the masses, elastic moduli, and uniaxial strengths of the specimens were increased. Denudation did not obviously occur on the specimen surface. There were more AE signals. At this time, the specimen fractures were mainly vertical. This is because the hydration reaction inside the specimens was not yet complete, and the solution immersion provided additional time for this reaction. Simultaneously, Cl^- fully reacted with the fly ash inside the specimen through diffusion and chemically activated the SiO_2 and Al_2O_3 in the fly ash, generating new gel products (C–S–H, C–S–H, AFt, etc.) to fill the internal fractures and pores in the specimens. The reaction process is given by Equations (9) and (10). The FTIR and XRD spectra revealed that AFt was not converted to Friedel's salt in large quantities, and AFt can provide higher strength to the specimens than Friedel's salt. This led to increases in the masses and uniaxial strengths of the specimens at the beginning of the dry–wet cycles. As indicated by the FE-SEM images, the mesostructures of the specimens became more compact, which enhanced both the mass and the uniaxial strength. Additionally, AE signals were generated by the refracture of the closed original fractures within the specimens under load.

$$Ca^{2+} + Al_2O_3 + 2OH^- \rightarrow CaO{\cdot}Al_2O_3{\cdot}H_2O \quad (9)$$

$$Ca^{2+} + Al_2O_3 + Cl^- + OH^- \rightarrow 3CaO{\cdot}Al_2O_3{\cdot}3CaCl_2{\cdot}10H_2O \quad (10)$$

3. At the 15th and 20th dry–wet cycles, the specimens' masses, elastic moduli, and uniaxial strengths were reduced. The denudation of the specimen surfaces was more severe, and the interlaced development of macroscopic vertical and horizontal fractures was observed. This is because, as the internal hydration reaction approached completion, the erosion failure of the specimens due to the chloride salt solution and the deterioration of the specimens due to the dry–wet cycles began to appear. First, the chloride salt entered the specimen's interior through diffusion and reacted chemically with the previously formed gel products to produce Friedel's salt. These products had certain swelling properties. The expansion and contraction of the gel products during the dry–wet cycles resulted in new fractures inside the specimens. The reaction process is given by Equation (11). Furthermore, the dry–wet cycles physically denudated the surfaces of the specimens through deterioration, and the internal structures of the specimens were also damaged. Under the coupled effects of these two factors and other related factors, the masses and uniaxial strengths of the specimens were reduced. The IR and XRD spectra revealed that Friedel's salt started to appear in the late stage of the dry–wet cycles. Additionally, the FE-SEM images

indicated that with the increase in the number of dry–wet cycles, the surfaces of the specimens started to deteriorate, and numerous pores and fractures appeared inside the specimen. As a result, the overall structure became loose.

$$3CaO{\cdot}Al_2O_3{\cdot}CaSO_4{\cdot}12H_2O + 2Cl^- \rightarrow 3CaO{\cdot}Al_2O_3{\cdot}CaCl_2{\cdot}10H_2O + SO_4^{2-} + 2H_2O \quad (11)$$

4. A higher-concentration solution can enter the specimen interior and enhance the uniaxial strength of the specimen faster owing to the higher concentration gradient. The strength deterioration rates of the specimens in the higher-concentration solution were lower than those for the low-concentration solution, but the surface denudation was significant. When the concentration of chloride salt in the solution is higher, the erosion on the surface of the specimen is more severe, and the chloride salt is more likely to invade the specimen interior and chemically react with the interior components, which has a positive effect on the specimen's strength. In a large amount, a solution with a high concentration of chloride salt can react with the internal components of the specimens or precipitate during the dry–wet cycles. This partially fills the fractures and pores inside the specimens and has a buffering effect on the strength deterioration of the specimens under the coupled effects of dry–wet cycles and chloride erosion. Moreover, a high concentration of Cl^- can maintain Friedel's salt; conversely, a lower concentration of Cl^- leads to a looser internal structure because it cannot maintain Friedel's salt, as shown in Equations (12) and (13). These factors led to the considerable strength deterioration of the specimens subjected to dry–wet cycles in the low-concentration solution. The IR and XRD spectra indicated that Friedel's salt started to appear late in the dry–wet cycles, and the FE-SEM images indicated that the specimens in the high-concentration solution had fewer pores and fractures than those in the low-concentration solution at the same number of dry–wet cycles. The specimens in the high-concentration solution had better integrity than those in the low-concentration solution.

$$CaCl_2 + 3Ca(OH)_2 + 12H_2O \rightarrow CaCl_2{\cdot}3Ca(OH)_2{\cdot}12H_2O \quad (12)$$

$$3CaO{\cdot}Al_2O_3{\cdot}CaCl_2{\cdot}10H_2O\ (s) \rightarrow 4Ca^{2+} + 2Al(OH)_4^- + 2Cl^- + 4OH^- + 4H_2O \quad (13)$$

6. Conclusions

1. At the early stage of dry–wet cycling in a chloride solution, the strengths of paste specimens were improved owing to the continued hydration reaction and chloride catalysis.
2. With increasing drying and wetting cycles of paste specimens in a chlorine salt solution, the erosion effect of chlorine salts and the deterioration effect of the dry–wet cycles began to appear, and the strengths of the paste specimens began to decline.
3. Compared with the low-concentration chlorine salt solution, the high-concentration solution can improve the strength of the paste faster and have a buffering effect on the strength deterioration of the paste.
4. The changes in the mechanical properties of desulphurised gypsum backfill samples under the coupled effects of chloride salt erosion and dry–wet cycles were investigated, and it was determined that the desulphurised gypsum backfill could maintain a certain strength in this special environment. The results provide a laboratory reference and theoretical basis for the field application of desulphurised gypsum backfill.
5. In this study, backfill samples were taken as the research objects. There are size differences between the samples and field backfill. Therefore, it is necessary to study the mechanical properties of backfill under the coupled effects of chloride erosion and dry–wet cycles at the field scale.

Author Contributions: Conceptualisation, S.W.; methodology, S.W.; software, S.W.; validation, S.W. and T.J.; data curation, D.Y.; writing—original draft preparation, S.W.; writing—review and editing, S.W., F.W., D.Y., T.J., and Z.Z.; funding acquisition, F.W. All authors have read and agreed to the published version of the manuscript.

Funding: This research was funded by the National Natural Science Foundation of China (Nos. 51904162, 51904167, and 52074169), the China Postdoctoral Science Foundation (No. 2020M682208), the SDUST Research Fund (No. 2019TDJH101), the Elite Talent Project of SDUST, and the Open Research Fund for the Key Laboratory of Safety and High-Efficiency Coal Mining (No. JYBSYS2019201).

Data Availability Statement: All data used to support the findings of this study are included within the article, and there are no restrictions on data access.

Conflicts of Interest: The authors declare no conflict of interest.

Abbreviations

FTIR	Fourier-Transform Infrared Spectroscopy
XRD	X-Ray Diffraction Analysis
FE-SEM	Field-Emission Scanning Electron Microscopy
XRF	X-Ray Fluorescence
Aft	Ettringite
C–A–H	$xCaO{\cdot}Al_2O_3{\cdot}yH_2O$
AE	Acoustic Emission
RA	The Ratio of Rise Time to Amplitude
PDS	Pressure Density Stage
LES	Linear Elastic Deformation Stage
PPYS	Pre-Peak Yielding Stage
PPDS	Post-Peak Damage Stage
C–S–H	$xCaO{\cdot}SiO_2{\cdot}yH_2O$
C_3SH	$3CaO{\cdot}SiO_2{\cdot}yH_2O$
C_4AH	$4CaO{\cdot}Al_2O_3{\cdot}H_2O$
C_3A	$3CaO{\cdot}Al_2O_3$
C_4AF	$3CaO{\cdot}Al_2O_3{\cdot}Fe_2O_3$

References

1. Wang, F.; Jiang, B.; Chen, S.; Ren, M. Surface collapse control under thick unconsolidated layers by backfilling strip mining in coal mines. *Int. J. Rock Mech. Min. Sci.* **2019**, *113*, 268–277. [CrossRef]
2. Li, L.; Yang, P. A numerical evaluation of continuous backfilling in cemented paste backfilled stope through an application of wick drains. *Int. J. Min. Sci. Technol.* **2015**, *25*, 897–904. [CrossRef]
3. Choudhary, B.; Kumar, S. Underground void filling by cemented mill tailings. *Int. J. Min. Sci. Technol.* **2013**, *23*, 893–900. [CrossRef]
4. Feng, G.; Jia, X.; Guo, Y.; Qi, T.; Li, D.; Li, Z.; Feng, J.; Liu, G.; Song, K.; Kang, L. Influence of the wasted concrete coarse aggregate on the performance of cemented paste backfill. *J. China Coal Soc.* **2015**, *40*, 1320–1325.
5. Chen, S.; Du, Z.; Zhang, Z.; Yin, D.; Feng, F.; Ma, J. Effects of red mud additions on gangue-cemented paste backfill properties. *Powder Technol.* **2020**, *367*, 833–840. [CrossRef]
6. Li, F.; Yin, D.; Zhu, C.; Wang, F.; Jiang, N.; Zhang, Z. Effects of Kaolin Addition on Mechanical Properties for Cemented Coal Gangue-Fly Ash Backfill under Uniaxial Loading. *Energies* **2021**, *14*, 3693. [CrossRef]
7. Cheng, H.-Y.; Wu, S.-C.; Zhang, X.-Q.; Wu, A.-X. Effect of particle gradation characteristics on yield stress of cemented paste backfill. *Int. J. Miner. Met. Mater.* **2020**, *27*, 10–17. [CrossRef]
8. Fall, M.; Pokharel, M. Coupled effects of sulphate and temperature on the strength development of cemented tailings backfills: Portland cement-paste backfill. *Cem. Concr. Compos.* **2010**, *32*, 819–828. [CrossRef]
9. Liu, H.; Zhang, J.; Zhou, N.; Sheng, L.; Zhang, L.; Zhu, L.; Wang, L. Study of the leaching and solidification mechanism of heavy metals from gangue-based cemented paste backfilling materials. *J. China Univ. Min. Technol.* **2021**, *50*, 523–531.
10. Zhang, J.; Tu, S.; Cao, Y.; Tan, Y.; Xin, H.; Pang, J. Coal gangue intelligent separation and backfilling technology and its engi-neering application in underground coal mine. *J. China Univ. Min. Technol.* **2021**, *50*, 523–531.
11. Chen, S.; Jiang, T.; Zhang, H.; Kong, K.; Bie, L. Emission reduction process for dechlorinating flue-gas desulfurization gypsum and reducing wastewater effluents: Application prospects from laboratory-scale studies. *Energy Sci. Eng.* **2020**, *8*, 2662–2679. [CrossRef]

12. Chandara, C.; Azizli, K.A.M.; Ahmad, Z.A.; Sakai, E. Use of waste gypsum to replace natural gypsum as set retarders in portland cement. *Waste Manag.* **2009**, *29*, 1675–1679. [CrossRef]
13. Hou, K. Flue gas desulfurization slag from power plant replaces natural gypsum as cement retarder. *Cement* **2005**, *10*, 21–23.
14. Tzouvalas, G.; Rantis, G.; Tsimas, S. Alternative calcium-sulfate-bearing materials as cement retarders: Part II. FGD gypsum. *Cem. Concr. Res.* **2004**, *34*, 2119–2125. [CrossRef]
15. Colak, A. The long-term durability performance of gypsum–Portland cement–natural pozzolan blends. *Cem. Concr. Res.* **2002**, *32*, 109–115. [CrossRef]
16. Guo, X.L.; Shi, H.S. Thermal treatment and utilization of flue gas desulphurization gypsum as an admixture in cement and concrete. *Constr. Build. Mater.* **2008**, *22*, 1471–1476. [CrossRef]
17. Guo, X.-L.; Shi, H.-S.; Liu, H.-Y. Effects of a combined admixture of slag powder and thermally treated flue gas desulphurization (FGD) gypsum on the compressive strength and durability of concrete. *Mater. Struct.* **2008**, *42*, 263–270. [CrossRef]
18. Zhang, Z.; Gao, L.; Yang, L.; Zhang, L.; Zhao, F. Study on water resistant gypsum blocks from desulfurization gypsum. *Fly Ash Compr. Utilizat.* **2009**, *2*, 27–30.
19. Qiao, X.; Poon, C.; Cheeseman, C. Use of flue gas desulphurization (FGD) waste and rejected fly ash in waste stabilization systems. *Waste Manag.* **2006**, *26*, 141–149. [CrossRef]
20. Li, B.; Yin, H.; Mao, X.; Li, Y.; Zhang, L.; Liu, R.; Qiu, P. Macroscopic and microscopic fracture features of concrete used in coal mine under chlorine salt erosion. *Int. J. Min. Sci. Technol.* **2016**, *26*, 455–459. [CrossRef]
21. Zhang, H. Effect of Chloride on Early Mechanical Properties of Cement Paste Filling Body. In Proceedings of the Technical Forum on Wastewater Treatment and Zero Discharge of High-salt Wastewater, Hangzhou, China, 25 October 2019.
22. Zhang, J.; Yao, Y.; Wang, X.; Guo, S.; Dong, G. Study on the combination of Ettringite and chloride ion in different chloride solutions. *J. Build. Mater.* **2021**, *6*, 1–9.
23. Ekolu, S.; Thomas, M.; Hooton, R.D. Pessimum effect of externally applied chlorides on expansion due to delayed ettringite formation: Proposed mechanism. *Cem. Concr. Res.* **2006**, *36*, 688–696. [CrossRef]
24. Du, Z.; Chen, S.; Yin, D.; Yao, D.; Zhang, Z. Experimental study of stability of paste backfill under chloride erosion environment. *J. China Univ. Min. Technol.* **2021**, *50*, 532–538, 547.
25. Zhu, Q.; Jiang, L.; Chen, Y.; Xu, J.; Mo, L. Effect of chloride salt type on chloride binding behavior of concrete. *Constr. Build. Mater.* **2012**, *37*, 512–517. [CrossRef]
26. Farnam, Y.; Washington, T.; Weiss, W. The Influence of Calcium Chloride Salt Solution on the Transport Properties, of Cementitious Materials. *Adv. Civ. Eng.* **2015**, *2015*, 1–13. [CrossRef]
27. Chen, S.; Du, Z.; Zhang, Z.; Zhang, H.; Xia, Z.; Feng, F. Effects of chloride on the early mechanical properties and microstructure of gangue-cemented paste backfill. *Constr. Build. Mater.* **2020**, *235*, 117504. [CrossRef]
28. Wang, X.; Shi, C.; He, F.; Yuan, Q.; Wang, D.; Huang, Y.; Li, Q. Chloride binding and its effects on microstructure of cement-based materials. *J. Chin. Ceram. Society* **2013**, *41*, 187–198.
29. Chen, S.; Jiang, T.; Wang, H.; Feng, F.; Yin, D.; Li, X. Influence of cyclic wetting-drying on the mechanical strength characteristics of coal samples: A laboratory-scale study. *Energy Sci. Eng.* **2019**, *7*, 3020–3037. [CrossRef]
30. Li, Y.; Ba, M.; Liu, J.; He, Z. Resistance to chloride erosion of cement matrix composite materials under dry-wet cycling and their micro-structural changes. *Acta Mater. Comp. Sin.* **2017**, *34*, 2856–2865.
31. Chen, Y.; Gao, J.; Tang, L.; Li, X. Resistance of concrete against combined attack of chloride and sulfate under drying–wetting cycles. *Constr. Build. Mater.* **2016**, *106*, 650–658. [CrossRef]
32. Aldaood, A.; Bouasker, M.; Al-Mukhtar, M. Impact of wetting–drying cycles on the microstructure and mechanical properties of lime-stabilized gypseous soils. *Eng. Geol.* **2014**, *174*, 11–21. [CrossRef]
33. Kalkan, E. Impact of wetting–drying cycles on swelling behavior of clayey soils modified by silica fume. *Appl. Clay Sci.* **2011**, *52*, 345–352. [CrossRef]
34. Zhou, S.; Lv, H.; Wu, Y. Degradation behavior of concrete under corrosive coal mine environment. *Int. J. Min. Sci. Technol.* **2019**, *29*, 307–312. [CrossRef]
35. Zhang, X.; Chen, R.; Jian, W. Study on water conversion law and solidification mechanism of cement-slag-fly ash solidified silt. *J. Eng. Geol.* **2021**, *6*, 1–11.
36. Jiang, T. Experimental Study on the Effect of Flue Gas Desulfurization Gypsum Dechlorinating on the Properties of Paste Back-Filling Materials. Master's Thesis, Shandong University of Science and Technology, Qingdao, China, 2021.
37. Ma, J.; Jiang, N.; Wang, X.; Jia, X.; Yao, D. Numerical Study of the Strength and Characteristics of Sandstone Samples with Combined Double Hole and Double Fissure Defects. *Sustainability* **2021**, *13*, 7090. [CrossRef]
38. Wang, Z. Effect of dry-wet cycles on mechanical properties of fly ash high-moisture materials. *Fly Ash Compr. Utilizat.* **2018**, *3*, 40–42.
39. Ammasi, A.K. Ragul Strength and Durability of High Volume Fly Ash in Engineered Cementitious Composites. *Mater. Today Proc.* **2018**, *5*, 24050–24058. [CrossRef]
40. Li, Z.; Ma, J. Preparation of Ca-Al hydrotalcite-like compounds and their desilication in sodium aluminate solution. *Sci. Sin. Chim.* **2010**, *40*, 577–584. [CrossRef]
41. Sun, R.; Hu, X.; Ling, Y.; Zuo, Z.; Zhuang, P.; Wang, F. Chloride diffusion behavior of engineered cementitious composite under dry-wet cycles. *Constr. Build. Mater.* **2020**, *260*, 119943. [CrossRef]

42. Hirao, H.; Yamada, K.; Takahashi, H.; Zibara, H. Chloride Binding of Cement Estimated by Binding Isotherms of Hydrates. *J. Adv. Concr. Technol.* **2005**, *3*, 77–84. [CrossRef]
43. Suryavanshi, A.; Narayan, S. Stability of Friedel's salt in carbonated concrete structural elements. *Cem. Concr. Res.* **1996**, *26*, 729–741. [CrossRef]
44. Wang, X.; Wang, E.; Liu, X.; Jia, H.; Li, D.; Wang, H. Time-frequency-space response of damage evolution for concrete based on AE-UT joint monitoring under step-loading. *Int. J. Min. Sci. Technol.* **2019**, *48*, 268–277.
45. Shiotani, T.; Ohtsu, M.; Ikeda, K. Detection and evaluation of AE waves due to rock deformation. *Constr. Build. Mater.* **2001**, *15*, 235–246. [CrossRef]
46. Lu, Q.; Wang, X.; Fu, S.; Li, Y. Effect of desulphurization gypsum on the strength and microstructure of the sludge ash cementitious system. *J. Zhengzhou Univ. (Eng. Sci.)* **2021**, *6*, 1–6.

minerals

MDPI

Article

Research on Strength Prediction Model and Microscopic Analysis of Mechanical Characteristics of Cemented Tailings Backfill under Fractal Theory

Hongwei Deng, Tao Duan, Guanglin Tian *, Yao Liu and Weiyou Zhang

School of Resources and Safety Engineering, Central South University, Changsha 410083, China; denghw208@126.com (H.D.); 205511029@csu.edu.cn (T.D.); 195512097@csu.edu.cn (Y.L.); hhzwy19950323@163.com (W.Z.)
* Correspondence: tgl15352006270@163.com

Abstract: In order to further study the internal relationship between the microscopic pore characteristics and macroscopic mechanical properties of cemented tailings backfill (CTB), in this study, mine tailings and ordinary Portland cement (PC32.5) were selected as aggregate and cementing materials, respectively, and different additives (anionic polyacrylamide (APAM), lime and fly ash) were added to backfill samples with mass concentration of 74% and cement–sand ratios of 1:4, 1:6 and 1:8. After 28 days of curing, based on the uniaxial compressive strength test, nuclear magnetic resonance (NMR) porosity test and the fractal characteristics of pore structure, the relationships of the compressive strength with the proportion and fractal dimension of pores with different radii were analyzed. The uniaxial compressive strength prediction model of the CTB with the proportion of harmless pores and the fractal dimension of harmful pores as independent variables was established. The results show that the internal pores of the material are mainly the harmless and less harmful pores, and the sum of the average proportions of the two reaches 73.45%. Some characterization parameters of pore structure have a high correlation with the compressive strength. Among them, the correlation coefficients of compressive strength with the proportion of harmless pores and fractal dimension of harmful pores are 0.9219 and 0.9049, respectively. The regression results of the strength prediction model are significant, and the correlation coefficient is 0.9524. The predicted strength value is close to the actual strength value, and the predicted results are accurate and reliable.

Keywords: backfill; compressive strength; fractal dimension; pore radius division; prediction model

Citation: Deng, H.; Duan, T.; Tian, G.; Liu, Y.; Zhang, W. Research on Strength Prediction Model and Microscopic Analysis of Mechanical Characteristics of Cemented Tailings Backfill under Fractal Theory. *Minerals* **2021**, *11*, 886. https://doi.org/10.3390/min11080886

Academic Editors: Chongchong Qi, Guichen Li, Hakan Basarir, Yuantian Sun and Qiusong Chen

Received: 10 July 2021
Accepted: 14 August 2021
Published: 16 August 2021

1. Introduction

Based on the background of vigorously advocating for the construction of "green mines", mining will always develop around the direction of "green, low carbon, safety and environmental protection". In order to meet this development keynote, a *Minerals* Special Issue recently focused on "Valorization of Metallurgical and Mining Residues and Wastes". Rybak et al. [1], considering artificial wastes instead of traditional backfill components, indicated that resource-reproducing technologies allow mining and processing enterprises to make full use of industrial mineral resources, resulting in multiple ecological and economic effects. Bagani et al. [2] discussed the possible use of nepheline syenite as a mine by-product in leaching alumina from azeotropic HCl solution. Tsaousi et al. [3] demonstrated that the industrial slag (copper slag) of the Swedish mining and smelting company "Boliden" was converted into advanced building materials through geological polymerization. Dentoni et al. [4] introduced the processing method of red mud in bauxite processing in the alumina industry. In addition, the filling mining method is highlighted in many mining methods by virtue of its unique advantages and has increasingly become the preferred method for the mining of metal and nonmetal underground mines [5]. At present, the strength of cemented backfill is one of the hot areas studied by scholars worldwide and is also an important index for judging the quality of backfill [6,7]. It has been found that the microscopic pore structure of backfill is closely related to its strength. The different pore microstructure characteristics of cemented backfill have an important impact on the mechanical properties of backfill. Therefore, it is of great practical significance to study the relationship between the microscopic pore structure of cemented backfill and the macroscopic mechanical properties to guide the actual production of mines.

Many scholars used different microscopic detection techniques to study the pore structure of backfill materials and established its relationship with macroscopic mechanical properties from different perspectives. Zhang [8] studied the effect of pore change on strength of cemented foam backfill (CFB) by mercury intrusion porosimetry (MIP). The results showed that the change of aggregate gradation and foaming agent dosage mainly affects the large pore volume of CFB and then affects the overall porosity of CFB, resulting in strength change. Deng [9] studied the variation of porosity and strength growth rate of three groups of CTB with different additives at different curing ages and established the corresponding strength evolution model of CTB. Liu [10] also studied the pore structure of backfill using NMR and scanning electron microscope (SEM) techniques and found that the pores of backfill were mainly small and harmless pores, also studying the internal relationship between the pore structure of backfill and unconfined compressive strength. Xin Jie [11] used digital image technology to quantify the pore characteristics of backfill materials and studied the relationship between the microstructure and mechanical properties of backfill materials. Yilmaz [12] analyzed the porosity of backfill materials of construction demolition waste (CDW) as a partial substitute for sulfide tailings by MIP and found that CDW can reduce the porosity to increase the strength of the backfills. Hu [13] studied the pore characteristics of backfill under four different humidity conditions through NMR and SEM techniques and established the relationship between the permeability of backfill materials and the uniaxial compressive strength. Yilmaz et al. [14] used an improved laboratory apparatus called CUAPS (pressure curing) and conventional plastic molds to study the effects of curing conditions on microstructure changes and corresponding unconfined compressive strength (UCS) of cemented paste backfill (CPB). Rubashkina et al. [15] added blast-furnace granular slag screenings 0-5 mm in size in the preparation of cemented filling mixture to optimize the gradation of low-quality fine sand and ultrafine sand. In addition, Chen Q [16–18] and Zhang Q [19] et al. also conducted in-depth research from the perspectives of environmental effects, rheological properties and so on of CPB, and achieved corresponding results.

In order to better describe the complexity and variability of the pore structure of porous materials, fractal theory has been applied to the study of the pore structure of porous materials in more and more research. Hu [20] studied the pore structure of stone powder CTB, quantified the fractal dimension of pores, calculated the grading index of mixed aggregate and studied the relationships of strength with grading index and pore structure. Deng [21] studied and analyzed the microscopic characteristics such as porosity and pore size distribution of sandstone-like materials and established a compressive strength prediction model with the proportion of macropores and the fractal dimension of macropores as independent variables. Zhang [22] studied the pore structure of electrically broken coal samples by using NMR and fractal theory and quantitatively studied the relationship between porosity growth and breakdown capacity. Chen Bowen [23] studied and analyzed the compressive strength characteristics of phosphogypsum cemented backfills by activity rate, particle size fractal dimension and pore fractal dimension.

In summary, the microscopic pore characteristics of backfill materials and the internal relationship between pore structure and strength of backfill materials have been tested by many scholars through different technical methods. In addition, with the in-depth study of rock materials, it has been found that the pore distribution of different internal radii in rock materials has a great influence on its mechanical properties [24,25]. However, it can be seen from the existing research results that there are still few studies on CTB in this regard. In view of this, the uniaxial compressive strength test and NMR porosity test were carried out on the backfill samples with different additives, and the internal relationships of the compressive strength of the CTB with the proportion and the fractal dimension of pores with different radii were analyzed. Based on the proportion of pores and fractal dimension of pores, the uniaxial compressive strength prediction model of the CTB was established.

2. Experiment Description

2.1. Experimental Materials and Scheme

In order to study the influence of different additives and different cement–sand ratios on the mechanical properties and pore structure of backfill materials, the aggregate selected in the experiment was tailings from a copper mine in Guangxi Province, China. Ordinary Portland cement (PC32.5) was selected as the cementing material for the experiment, and different additives (APAM, lime and fly ash) were added. Four groups of backfill samples with mass concentration of 74% and cement–sand ratios of 1:4, 1:6 and 1:8 were prepared: group A (only adding cement), group B (APAM (1% of cement content) and cement), group C (lime (5% of cement content) and cement) and group D (fly ash (10% of cement content) and cement). The experimental water was tap water in the laboratory. The basic physical parameters of tailings are shown in Table 1. The main chemical components of raw materials are shown in Table 2. The particle size distribution and cumulative curve are shown in Figure 1.

Table 1. Basic physical parameters of tailings (source: elaborated by authors).

Tailing Type	Specific Weight	Loose Density ($t \times m^{-3}$)	Tap Density ($t \times m^{-3}$)	Maximum Porosity (%)	Minimum Porosity (%)	Natural Repose Angle (°)
Graded tailings	2.727	1.466	1.702	0.462	0.376	39

Table 2. Main chemical compositions of the materials (source: elaborated by authors).

Material	Tailing	Cement	Lime	Fly Ash
Main Component	SiO_2 (79.27%) CaO (5.54%) Fe_2O_3 (5.34%) Al_2O_3 (4.16%) SO_3 (2.62%) K_2O (0.82%)	CaO (63.34%) SiO_2 (23.53%) Al_2O_3 (4.23%) Fe_2O_3 (3.45%) MgO (2.83%) Na_2O (0.03%)	CaO (73.56%) SiO_2 (2.14%) MgO (2.14%) Al_2O_3 (1.47%) Fe_2O_3 (0.98%) K_2O (0.23%)	SiO_2 (46.31%) Al_2O_3 (28.45%) CaO (9.24%) Fe_2O_3 (4.25%) Ti_2O (1.30%) SO_3 (0.85%)

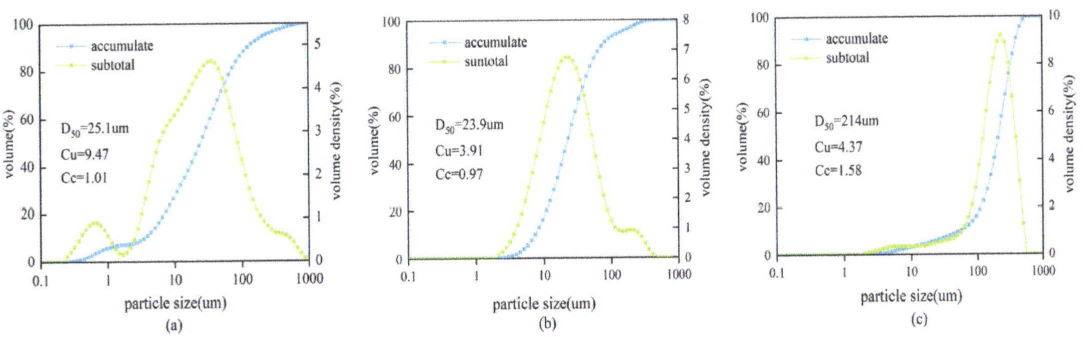

Figure 1. Material particle size distribution and accumulation curve: (**a**) fly ash; (**b**) cement; (**c**) tailings (source: elaborated by authors).

2.2. Sample Preparation and Test

According to the above experimental scheme, combined with the specific sample size requirements (cylinder with diameter of 50 mm × height of 100 mm) of "Rock Test Procedures for Water Conservancy and Hydropower Engineering" [26] and "Test Specification for Hydraulic Concrete" [27], the experimental preparation steps and experimental tests of the samples were as follows:

(1) Pre-preparation: The recovered tailings were naturally air-dried for 1 to 2 days. After evaporation of water on the surface of the tailings, the tailings were put into the drying box for drying. This step also includes screening impurities, mold cleaning and other work.

(2) Stirring into the mold: The experimental raw materials were weighed well and fully stirred for about 5 min. After the filling slurry was uniform, and no bubbles were generated; it could be put into the mold with lubricating oil on the inner wall and stabbed evenly with a glass rod.

(3) Demolding and maintenance: The filling slurry with mold was maintained for 1 to 2 days in the laboratory. After the initial setting of the filling slurry with a certain strength that can maintain self-support, the backfill samples were demolded and numbered and finally put into the concrete standard curing box with 20 °C temperature and 99% relative humidity for 28 days.

(4) Experimental test: The size, weight and acoustic wave velocity of the curing backfill samples were measured. AniMR-150 rock NMR analysis system (NIUMAG, Suzhou, China) was used to obtain the porosity of backfill samples and pore distribution parameters of different internal radii. Before the porosity test, in order to fill the pores in the backfill samples with water, the rock vacuum saturation device was used for saturation treatment. In addition, in order to ensure that the backfill samples were fully saturated, it was necessary to soak them in distilled water for about 1 h, and the saturated backfill samples were wrapped with preservative film to test their porosity. In order to obtain the uniaxial compressive strength of backfill samples, SHT4206 microcomputer-controlled electro-hydraulic servo universal material testing machine (SANS, Shanghai, China) was used in the test, and the backfill samples were pressurized by force control, with the loading speed of 100 N/s. The specific experimental process is shown in Figure 2.

2.3. Pore Radius Division and Fractal Dimension Calculation

According to the basic principle of NMR [28], the surface relaxation of pore water inside the backfill can be expressed by Formula (1) as follows:

$$\frac{1}{T_2} = \rho_2 \frac{S}{V} \tag{1}$$

In Formula (1), ρ_2 is surface relaxation strength. S/V is the ratio of pore surface area to fluid volume.

Since the pore radius is proportional to the pore throat diameter [29], the formula (1) can be simplified to:

$$\frac{1}{T_2} = \rho_2 \frac{F_s}{R} \tag{2}$$

In Formula (2), F_s is the geometric factor (for cylindrical pores, F_s is 2) and R is the pore radius. Due to the lack of corresponding research on ρ_2, a relatively accurate value cannot be obtained. Considering that the mineral and material components of the backfill are similar to those of concrete cement, according to the study in [30], the surface relaxation rate ρ_2 of the backfill can be selected as 12 nm/ms. Therefore, Formula (2) can be simplified as:

$$R = 24T_2 \tag{3}$$

According to formula (3), the relationship between pore radius and relaxation time is a one-to-one correspondence; that is, the T_2 value obtained by NMR test can be converted into the pore radius of the backfill.

According to the existing research results [31], the pore radius of backfill can be divided into four categories: harmless pores (pore radius < 20nm, a type of pores having no effect on strength), less harmful pores (20 nm < pore radius < 100 nm, a type of pores having little effect on strength), harmful pores (100 nm < pore radius < 200 nm, a type of pores having effect on strength) and more harmful pores (pore radius > 200 nm, a type of pores having great influence on strength). Combined with the relaxation time of NMR detection, the pore radius in the material was divided into harmless pores, less harmful pores, harmful pores and more harmful pores. The pore radius division is shown in Figure 3.

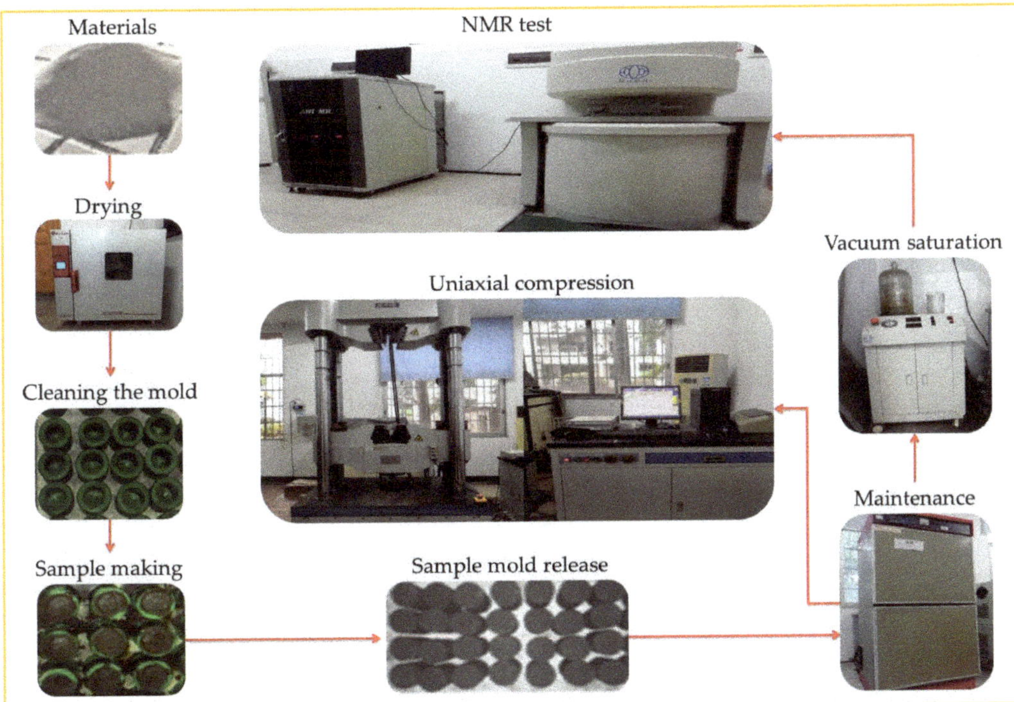

Figure 2. Experimental steps (source: elaborated by authors).

The proportion of pores with different radii can be expressed as:

$$P_i = \frac{V_r}{V_t} \times 100\% = \frac{V_i - V_{i-1}}{V_4} \times 100\% \ (i = 1, 2, 3, 4) \quad (4)$$

In Formula (4), V_r is the cumulative volume fraction of pores smaller than r, V_t is the total porosity, $V_0 = 0$, V_1 is the cumulative volume fraction of pores smaller than 20 nm, V_2 is the cumulative volume fraction of pores smaller than 100 nm, V_3 is the cumulative volume fraction of pores smaller than 200 nm and V_4 is the total porosity. The calculation results are shown in Table 3.

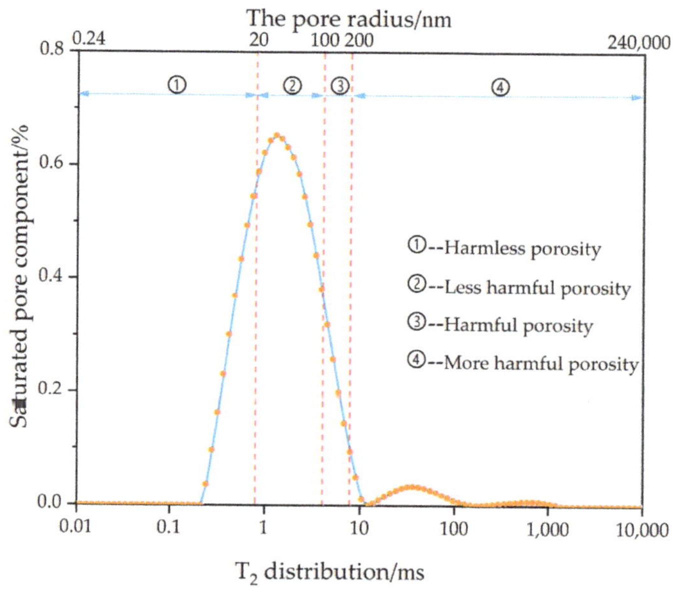

Figure 3. Pore radius division of backfill (source: elaborated by authors).

Table 3. Table of pore radius distribution and strength of backfill samples (source: elaborated by authors).

Group	Cement–Sand Ratio	Porosity (%)	Proportion of Harmless Pores (%)	Proportion of Less Harmful Pores (%)	Proportion of Harmful Pores (%)	Proportion of More Harmful Pores (%)	Uniaxial Compressive Strength (MPa)
Group A	1:4	12.280	29.58	56.77	9.69	3.96	2.513
		8.509	59.51	38.07	0.13	2.29	5.245
		15.539	18.43	52.80	14.58	14.19	1.919
	1:6	10.357	25.11	57.08	13.45	4.36	3.244
		16.714	13.11	44.83	14.21	27.85	1.486
		18.291	4.47	50.49	23.13	21.91	1.105
	1:8	15.431	18.76	59.36	14.78	7.10	1.936
		17.928	6.72	51.87	21.58	19.83	1.146
		21.578	2.33	53.09	26.38	18.20	0.942
Group B	1:4	10.261	43.51	43.94	4.74	7.81	3.623
		13.198	37.52	50.25	4.88	7.35	2.327
		17.539	9.05	54.33	20.91	15.71	1.247
	1:6	16.377	14.54	50.90	17.16	17.40	1.730
		11.084	35.42	48.76	6.93	8.89	2.539
		16.463	13.93	52.40	18.95	14.72	1.664
	1:8	21.361	3.21	54.05	26.37	16.37	1.010
		10.277	23.14	53.18	12.96	10.72	3.552
		19.446	3.24	56.87	23.95	15.94	1.045
Group C	1:4	10.540	32.35	54.47	6.28	6.90	3.101
		12.328	23.52	46.52	11.91	18.05	2.376
		12.762	28.87	54.35	9.58	7.20	2.465
	1:6	15.734	14.91	56.03	15.05	14.01	1.831
		10.191	50.09	45.87	2.36	1.68	4.792
		18.635	5.79	54.23	18.88	21.10	1.102
	1:8	16.845	11.60	60.85	15.75	11.80	1.316
		17.623	15.83	57.36	13.71	13.10	1.342
		15.483	24.32	55.43	12.21	8.04	1.166

Table 3. *Cont.*

Group	Cement–Sand Ratio	Porosity (%)	Proportion of Harmless Pores (%)	Proportion of Less Harmful Pores (%)	Proportion of Harmful Pores (%)	Proportion of More Harmful Pores (%)	Uniaxial Compressive Strength (MPa)
Group D	1:4	11.036 14.153 10.234	27.33 22.39 56.53	47.61 56.03 36.34	14.65 10.10 0.98	10.41 11.48 6.15	2.647 2.323 4.322
	1:6	11.024 14.762 17.843	25.00 21.70 7.57	50.20 60.12 57.12	11.20 11.52 19.12	13.60 6.66 16.19	2.980 1.952 1.190
	1:8	15.669 18.852 18.259	18.32 3.46 4.10	63.23 45.62 58.36	12.70 22.33 24.65	5.75 28.59 12.89	1.899 1.054 1.126

In recent years, in order to fully describe and show the variability and complexity of pore structure of backfill, many experts and scholars have introduced fractal theory to characterize the complex microscopic pore structure inside the material. According to the existing research results of fractal theory [32,33], T_2 relaxation time distribution and fractal dimension can be expressed as follows:

$$\lg(V) = (3 - D)\lg T_2 + (D - 3)\lg T_{2max} \quad (5)$$

In Formula (5), V is the cumulative volume fraction of pores smaller than r, and D is the fractal dimension. If the pores of backfill have self-similar structure and fractal characteristics, there is a linear relationship between lg(V) and $\lg T_2$. Combined with the above classification basis of pores with different radii, the fractal dimension corresponding to different pores can be calculated. The calculation results are shown in Table 4.

Table 4. Fractal dimension of backfill samples (source: elaborated by authors).

Group	Cement–Sand Ratio	Less Harmful Pores		Harmful Pores		More Harmful Pores	
		D	R^2	D	R^2	D	R^2
Group A	1:4	2.3467 2.7199 2.2163	0.9678 0.9100 0.9626	2.8361 2.9987 2.7795	0.9649 0.6203 0.9777	2.9793 2.9976 2.9829	0.8767 0.6508 0.7553
	1:6	2.4976 2.1541 1.6254	0.9520 0.9564 0.9532	2.9073 2.7403 2.5831	0.9583 0.9904 0.9841	2.9878 2.9644 2.9773	0.8106 0.6861 0.6075
	1:8	2.1656 1.7197 1.5054	0.9655 0.9569 0.9585	2.7951 2.6280 2.5380	0.9574 0.9796 0.9804	2.9898 2.9765 2.9819	0.8799 0.7502 0.6950
Group B	1:4	2.6008 2.2989 1.8907	0.9430 0.9659 0.9619	2.9384 2.8145 2.6616	0.9455 0.9709 0.9807	2.9889 2.9857 2.9841	0.8755 0.8304 0.6421
	1:6	2.1328 2.2949 1.9667	0.9641 0.9699 0.9614	2.7238 2.8205 2.7020	0.9857 0.9582 0.9771	2.9826 2.9949 2.9832	0.5916 0.6776 0.7415
	1:8	1.4423 2.5112 1.6144	0.9490 0.9470 0.9564	2.5502 2.9373 2.6026	0.9812 0.8995 0.9771	2.9856 2.9891 2.9835	0.5668 0.8378 0.7184

Table 4. *Cont.*

Group	Cement–Sand Ratio	Less Harmful Pores		Harmful Pores		More Harmful Pores	
		D	R^2	D	R^2	D	R^2
Group C	1:4	2.4294 2.3599 2.2637	0.9514 0.9641 0.9106	2.9186 2.8141 2.8034	0.9045 0.9819 0.9716	2.9896 2.9746 2.9891	0.8226 0.8308 0.7439
	1:6	2.1091 2.7270 1.5193	0.9552 0.9052 0.9483	2.7726 2.9896 2.6055	0.9758 0.9104 0.9762	2.9843 2.9937 2.9881	0.6803 0.6295 0.6065
	1:8	1.9678 1.7234 1.6531	0.9509 0.9539 0.9362	2.7679 2.5864 2.5736	0.9654 0.9868 0.9876	2.9856 2.9631 2.9735	0.7453 0.6467 0.6013
Group D	1:4	2.3966 2.2765 2.6091	0.9742 0.9536 0.9324	2.7879 2.8578 2.9721	0.9847 0.9534 0.8073	2.9903 2.9840 2.9935	0.6127 0.8327 0.6249
	1:6	2.3680 2.2299 1.7933	0.9643 0.9598 0.9546	2.8747 2.8449 2.6935	0.9271 0.9389 0.9704	2.9943 2.9893 2.9785	0.7766 0.8936 0.8181
	1:8	2.1423 1.5580 1.7147	0.9543 0.9514 0.9495	2.8295 2.5547 2.6764	0.9449 0.9860 0.9744	2.9925 2.9647 2.9707	0.6983 0.7478 0.8522

3. Analysis of Experimental Results

3.1. Analysis of Pore Structure Characterization Parameters

Table 3 shows that the porosity of the backfill material ranges from 8.509 to 21.578%, and the variation range is relatively large, which is mainly related to the previous sample preparation process. In general, the internal pores of the backfill material are mainly the harmless pores and less harmful pores, and the sum of the average proportions of the two reaches 73.45%. However, there are fewer harmful pores and more harmful pores, and the sum of their average proportions is only 26.55%. Among them, the proportion of less harmful pores varies from 36.34 to 63.23%, and the average proportion of pores is the largest, reaching 52.47%. The second is the average proportion of harmless pores, which is 20.98%, and the proportion ranges from 2.33 to 59.51%. However, the proportions of harmful pores and more harmful pores are relatively small, and the average proportions of the two pore types are 14.10% and 12.45%, respectively. The variation ranges of the proportions of the two pore types are 0.13–26.38% and 1.68–28.59%, respectively.

The above analysis of pore distribution of backfill samples has similar results to the published related research. Liu [10] found that the pores of CPB samples are mainly non-damaging pores, and the total volume proportion of "less-damaging" pores and "non-damaging" pores is more than 80%. In addition, Ruidong Wu [34] studied the change of pore structure before and after freeze–thaw of iron tailings powder concrete. It was also found that before freeze–thaw of concrete samples, the porosity of the samples was mainly composed of harmless pores and less harmful pores, while there were fewer harmful pores and more harmful pores. Similarly, in [35], when studying the effect of fly ash with different particle sizes on concrete, it was found that among the pores of concrete samples with different particle sizes and fly ash contents, the porosity of the samples was mainly composed of harmless pores and less harmful pores.

Table 4 shows that in the pore fractal dimension of the backfill materials, the fractal dimension calculated by curve fitting is very low due to the pore division of harmless pores being too small (<20 nm) and the corresponding transverse relaxation time T_2 being less than 1 ms, so the fractal dimension of harmless pores is not considered. It is also mentioned in [10] that "when the pore radius was smaller than 20nm or larger than 1um, the fitting goodness was very poor, indicating that within these size ranges, the pores did not have fractal features". The fractal dimension analysis results of the other three types of pores are as follows: The fractal dimension of less harmful pores ranges from 1.4423 to 2.7270, with an average of 2.0985, and the correlation coefficient of their fractal dimension ranges from 0.9052 to 0.9742, with an average of 0.9520, indicating that the fractal dimension of less harmful pores has high reliability. The second is the fractal dimension of harmful pores, whose fractal dimension ranges from 2.5380 to 2.9987 with an average value of 2.7633, and their average correlation coefficient is 0.9510 with a range of 0.6203 to 0.9904, which is also reliable. The correlation coefficient of the fractal dimension of more harmful pores ranges from 0.5668 to 0.8936, with an average of 0.7322, and is lower than that of the first two types of pores, but their fractal dimension ranges from 2.9631 to 2.9976, with an average of 2.9837. From the above analysis, three types of pores in the backfill material all have obvious fractal characteristics.

3.2. *Correlation Analysis between Pore Structure Characterization Parameters and Compressive Strength*

In order to analyze the relationships of the compressive strength of the backfill material with the porosity, the proportion of different pores and the fractal dimension of different pores, the calculation results of the correlation coefficients are shown in Table 5.

Table 5. Table of correlation coefficients (source: elaborated by authors).

Correlation Coefficient	Porosity	Proportion of Harmless Pores	Proportion of Less Harmful Pores	Proportion of Harmful Pores	Proportion of More Harmful Pores	Fractal Dimension of Less Harmful Pores	Fractal Dimension of Harmful Pores	Fractal Dimension of More Harmful Pores
Uniaxial Compressive Strength	−0.9086	0.9219	−0.5655	−0.8477	−0.6808	0.9062	0.9049	0.6283

Table 5 shows that the absolute values of correlation coefficients of the compressive strength of the backfill material with the porosity and the proportion of different pores are all greater than 0.5655. Among them, the correlation coefficient between the proportion of harmless pores and the compressive strength is the highest, reaching up to 0.9219, and the correlation is positive, indicating that the proportion of harmless pores has a high correlation with the compressive strength. The second is the proportion of the porosity and the proportion of harmful pores, and the corresponding correlation coefficients are 0.9086 and 0.8477, respectively, with negative correlation. However, the proportion of less harmful pores and the proportion of harmful pores are not highly correlated with the compressive strength, and both of them are negatively correlated. In addition, the correlation coefficients of the compressive strength with the fractal dimension of less harmful pores, the fractal dimension of harmful pores and the fractal dimension of more harmful pores are all greater than 0.6283, and they are all positively correlated. Among them, the fractal dimension of less harmful pores and the fractal dimension of harmful pores have large correlation coefficients with the compressive strength, which are 0.9062 and 0.9049, respectively, indicating that they have a high correlation with the compressive strength. However, the correlation coefficient between the fractal dimension of more harmful pores and the compressive strength is lower than that of the first two types of pores, which is only 0.6283. The analysis results of the relationship between the fractal dimensions of pores with different radii and the compressive strength are consistent with the conclusion in [21].

4. Strength Prediction Model

4.1. *Selection of Model Parameters*

The selection of accurate and reasonable parameters will have a direct impact on the reliability and simplicity of the compressive strength prediction model. Based on the analysis of the above experimental results, the characterization methods of the microscopic pore structure of the backfill are diverse. In order to avoid the complexity of the model due to the selection of too many independent variables, the single-factor variance analysis of the pore structure characterization parameters and the compressive strength is carried out. The analysis results are shown in Table 6.

Table 6. Single-factor variance analysis of pore structure parameters and compressive strength (source: elaborated by authors).

Parameter Types	Pore Structure Parameters	R^2	F	P
Porosity and Proportions	Porosity	0.8256	160.9189	0.0000
	Proportion of Harmless Pores	0.8500	192.6101	0.0000
	Proportion of Less Harmful Pores	0.3198	15.9866	0.0003
	Proportion of Harmful Pores	0.7186	86.8409	0.0000
	Proportion of More Harmful Pores	0.4635	29.3793	0.0000
Fractal Dimension	Fractal Dimension of Less Harmful Pores	0.8212	156.1936	0.0000
	Fractal Dimension of Harmful Pores	0.8189	153.7063	0.0000
	Fractal Dimension of More Harmful Pores	0.3947	22.1745	0.0000

Table 6 shows that the correlation coefficients of the compressive strength with the porosity and the proportion of different pores are quite different. Among them, the porosity and the proportion of harmless pores are highly correlated with the compressive strength, and the correlation coefficients are 0.8256 and 0.8500, respectively. The correlation coefficient between the proportion of harmful pores and the compressive strength is 0.7186, which is the next highest. However, the proportion of less harmful pores and the proportion of more harmful pores have low correlations with the compressive strength. In addition, the fractal dimension of less harmful pores and the fractal dimension of harmful pores have a high correlation with the compressive strength, and the correlation coefficients are 0.8212 and 0.8189, respectively. However, the correlation between the fractal dimension of more harmful pores and the compressive strength is low. All of the test coefficient P values of the pore structure characterization parameters and the compressive strength are less than 0.05, indicating that the pore structure characterization parameters have a significant relationship with the compressive strength.

The analysis results of Table 6 show that the proportion of harmless pores and the fractal dimension of harmful pores are highly correlated with the compressive strength. Therefore, based on the above analysis, it is reasonable to select the proportion of harmless pores and the fractal dimension of harmful pores as the independent variables of the strength prediction model.

4.2. Establishment of Strength Prediction Model

The strength prediction model of the backfill is established by the proportion of harmless pores and the fractal dimension of harmful pores selected from the above analysis. The linear fitting relationships of the compressive strength with the proportion of harmless pores and the fractal dimension of harmful pores are obtained. The results are shown in Figures 4 and 5. It can be seen from the two figures that the proportion of harmless pores and the fractal dimension of harmful pores are positively linearly correlated with the compressive strength, and the correlation coefficients are 0.8500 and 0.8189, respectively. This shows that the linear correlation between the two pore structure characterization parameters and the compressive strength is very good.

Based on the above analysis, the strength prediction model to be established is a quadratic function of the two independent variables of the proportion of harmless pores and the fractal dimension of harmful pores. By repeatedly adjusting the expression of the model and referring to the relevant literature [21], the preset strength prediction model is expressed as:

$$M_c = \beta_0 + \beta_1 R_1 + \beta_2 D_2 + \beta_3 R_1 D_2 \quad (6)$$

In Formula (6), M_c is the uniaxial compressive strength of the materials, R_1 is the proportion of harmless pores and D_2 is the fractal dimension of harmful pores. The calculation results of multiple regression analysis of the above expression are shown in Table 7.

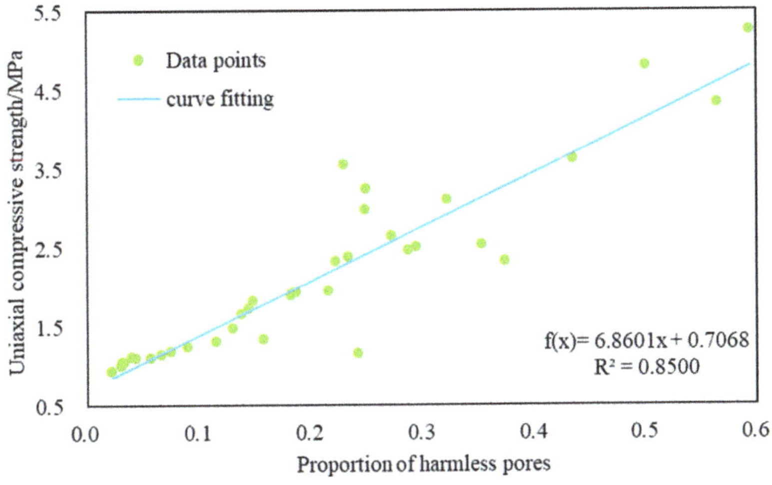

Figure 4. The fitting relationship between the proportion of harmless pores and strength (source: elaborated by authors).

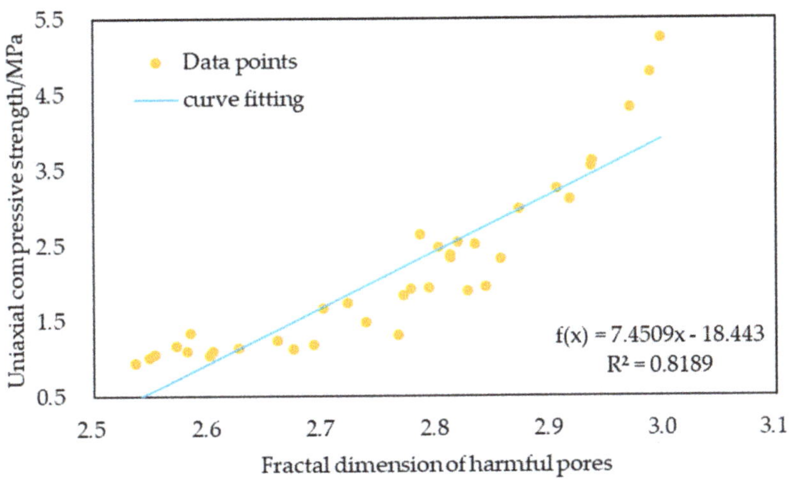

Figure 5. Fitting relationship between fractal dimension of harmful pores and strength (source: elaborated by authors).

Table 7. Calculation results of multiple regression analysis (source: elaborated by authors).

Regression Coefficients	Estimated Value	Confidence Interval
β_0	−5.0708	[−8.4638,−1.6778]
β_1	−30.9799	[−44.3521,−17.6077]
β_2	2.3595	[1.0801,3.6390]
β_3	12.0689	[7.4814,16.6564]
$R^2 = 0.9524$, $F = 213.5310$, $P = 0.0000$		

Table 7 shows that the correlation coefficient of the strength prediction model is 0.9524, which indicates that the multiple regression equation has a good fitting degree and the degree of closeness to the real values is also very high. In addition, the P value of the equation is far less than the given significance level of 0.05, indicating that the regression results of the strength prediction model are significant; that is, the relationships of the compressive strength with the proportion of harmless pores and the fractal dimension of harmful pores are significant. Substituting the estimated values of the regression coefficients calculated in Table 7 into the Formula (6), the equation of the backfill material strength prediction model can be expressed as follows:

$$M_c = -5.0708 - 30.9799R_1 + 2.3595D_2 + 12.0689R_1D_2 \quad (7)$$

4.3. Model Rationality Verification

In order to verify the accuracy of the strength prediction model, this paper uses the prediction model to calculate the uniaxial compressive strength of some backfill samples and compares it with the uniaxial compressive strength measured by the actual test. The analysis results are shown in Table 8 and Figure 6.

Table 8. Verification results of backfill strength prediction model (source: elaborated by authors).

Serial Number	Proportion of Harmless Pores (%)	Fractal Dimension of Harmful Pores	Measured Strength (MPa)	Predicted Strength (MPa)	Error (%)
1	46.07	2.7589	2.489	2.506	0.69
2	23.15	2.6625	1.483	1.478	0.31
3	65.70	2.6936	2.329	2.289	1.71
4	27.31	2.5769	1.021	1.042	2.09
5	39.82	2.6785	1.794	1.785	0.48
6	23.46	2.5871	1.104	1.091	1.21

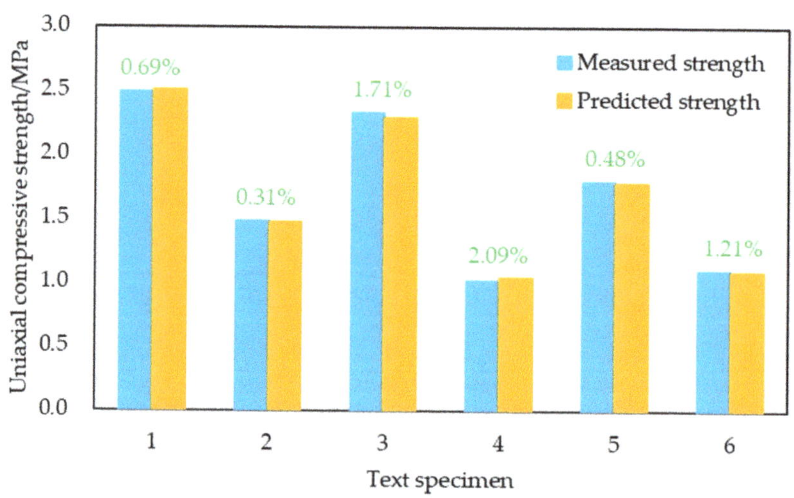

Figure 6. Comparison results of measured strength and predicted strength of backfill (source: elaborated by authors).

It can be seen from Table 8 and Figure 6 that the strength value calculated by the strength prediction model is very close to the strength value measured by the test, and the error is small, indicating that the established strength prediction model of the backfill material has high accuracy and reliability and can thus be used to predict the uniaxial compressive strength of relevant backfill materials.

5. Conclusions

In the experiment, four groups of backfill samples with different cement–sand ratios were made with raw materials. Based on uniaxial compressive strength test, NMR porosity test and fractal characteristics of pore structure, a compressive strength prediction model with the proportion of harmless pores and the fractal dimension of harmful pores as independent variables was established. The experimental conclusions are as follows:

(1) The internal pore distributions of the backfill material are mainly the harmless and less harmful pores, while the harmful and more harmful pores are less prevalent. The fractal dimensions of the three types of pores have obvious fractal characteristics. Among them, the correlation coefficient of the fractal dimension of less harmful and harmful pores is higher, while that of more harmful pores is lower than that of the first two types of pores.

(2) The porosity, the proportion of harmless pores, the proportion of harmful pores, the fractal dimension of less harmful pores and the fractal dimension of harmful pores all have a high correlation with the compressive strength. Therefore, it is reasonable to select the proportion of harmless pores and the fractal dimension of harmful pores as the independent variables of the strength prediction model.

(3) The compressive strength has strong linear relationships with the proportion of harmless pores and the fractal dimension of harmful pores, and they are positively linearly correlated.

(4) The regression results of the backfill strength prediction model established in this paper are very significant. The strength value calculated by the established prediction model is very close to the actual strength value measured by the test, and the error is small, indicating that the established model is reliable.

Author Contributions: H.D. and G.T. provided experimental ideas; Y.L. and T.D. carried out the experiment; W.Z. and G.T. directed the experiment; T.D. wrote the paper. All authors have read and agreed to the published version of the manuscript.

Funding: This work was financially supported by the National Natural Science Foundation of China under Grant number 51874352.

Data Availability Statement: The source data can be obtained in the article.

Acknowledgments: Not applicable.

Conflicts of Interest: The authors declare no conflict of interest.

References

1. Rybak, J.; Kongar-Syuryun, C.; Tyulyaeva, Y.; Khayrutdinov, A.M. Creation of Backfill Materials Based on Industrial Waste. *Minerals* **2021**, *11*, 739. [CrossRef]
2. Bagani, M.; Balomenos, E.; Panias, D. Nepheline Syenite as an Alternative Source for Aluminum Production. *Minerals* **2021**, *11*, 734. [CrossRef]
3. Tsaousi, G.-M.; Panias, D. Production, Properties and Performance of Slag-Based, Geopolymer Foams. *Minerals* **2021**, *11*, 732. [CrossRef]
4. Dentoni, V.; Grosso, B.; Pinna, F. Experimental Evaluation of PM Emission from Red Mud Basins Exposed to Wind Erosion. *Minerals* **2021**, *11*, 405. [CrossRef]
5. Zhang, J.; Ma, C.; Jin, A.; Chen, S. Experimental study on strength characteristics of cemented backfill. *Min. Res. Dev.* **2021**, *41*, 124–129.
6. Liu, T. *Technology and Application of Filling Mining*; Metallurgical Industry Press: Beijing, China, 2001.
7. Liu, C.; Han, B.; Sun, W.; Wu, J.; Yao, S.; Hu, H. Experimental study on strength characteristics of waste rock crushing cemented backfill in alpine area and its industrial application. *Chin. J. Rock Mech. Eng.* **2015**, *34*, 139–147.
8. Zhang, S.; Yang, L.; Qiu, J.; Hou, C.; Guo, Z. Fluidity and strength behaviors of cemented foam backfill: Effect of particle size distribution and foaming agent dosage. *Bull. Eng. Geol. Environ.* **2021**, *80*, 3177–3191. [CrossRef]
9. Deng, H.; Liu, Y.; Zhang, W.; Yu, S.; Tian, G. Study on the Strength Evolution Characteristics of Cemented Tailings Backfill from the Perspective of Porosity. *Minerals* **2021**, *11*, 82. [CrossRef]
10. Liu, L.; Fang, Z.; Qi, C.; Zhang, B.; Guo, L.; Song, K.-I. Experimental investigation on the relationship between pore characteristics and unconfined compressive strength of cemented paste backfill. *Constr. Build. Mater.* **2018**, *179*, 254–264. [CrossRef]

11. Xin, J. *Study on Microstructure and Mechanical Properties of Cemented Tailings Filling Material*; Xi'an University of Science and Technology: Xi'an, China, 2020.
12. Yilmaz, T.; Ercikdi, B.; Deveci, H. Utilisation of construction and demolition waste as cemented paste backfill material for underground mine openings. *J. Environ. Manag.* **2018**, *222*, 250–259. [CrossRef] [PubMed]
13. Jianhua, H.U.; Zhao, F.; Ren, Q.; Kuang, Y.; Zhou, T.; Luo, Z. Microscopic characterization and strength characteristics of cemented backfill under different humidity curing conditions. *R. Soc. Open Sci.* **2019**, *6*, 191227.
14. Yilmaz, E.; Belem, T.; Bussière, B.; Benzaazoua, M. Relationships between microstructural properties and compressive strength of consolidated and unconsolidated cemented paste backfills. *Cem. Concr. Compos.* **2011**, *33*, 702–715. [CrossRef]
15. Rubashkina, I.T.; Korneichuk, M.A. Optimization of grading of sand in backfill using metallurgical waste. *J. Min. Sci.* **2021**, *56*, 130–139.
16. Chen, Q.; Tao, Y.; Zhang, Q.; Qi, C. The rheological, mechanical and heavy metal leaching properties of cemented paste backfill under the influence of anionic polyacrylamide. *Chemosphere* **2021**, *286*, 131630. [CrossRef]
17. Chen, Q.; Sun, S.; Liu, Y.; Qi, C.; Zhou, H.; Zhang, Q. Immobilization and leaching characteristics of fluoride from phosphogypsum-based cemented paste backfill. *Int. J. Miner. Metall. Mater.* **2021**. [CrossRef]
18. Chen, Q.; Tao, Y.; Feng, Y.; Zhang, Q.; Liu, Y. Utilization of modified copper slag activated by Na2SO4 and CaO for unclassified lead/zinc mine tailings based cemented paste backfill. *J. Environ. Manag.* **2021**, *290*, 112608. [CrossRef]
19. Zhang, Q.; Li, Y.; Chen, Q.; Liu, Y.; Feng, Y.; Wang, D. Effects of temperatures and pH values on rheological properties of cemented paste backfill. *J. Cent. South Univ.* **2021**, *28*, 1707–1723. [CrossRef]
20. HU, J.-H.; Ren, Q.-F.; Yang, D.-J.; Ma, S.-W.; Shang, J.-L.; Ding, X.-T.; Luo, Z.-Q. Cross-scale characteristics of backfill material using NMR and fractal theory. *Trans. Nonferr. Met. Soc. China* **2020**, *30*, 1347–1363. [CrossRef]
21. Deng, H.; Tian, G.; Yu, S.; Jiang, Z.; Zhong, Z.; Zhang, Y. Research on Strength Prediction Model of Sand-Like Material Based on Nuclear Magnetic Resonance and Fractal Theory. *Appl. Sci.* **2020**, *10*, 6601. [CrossRef]
22. Zhang, X.; Lin, B.; Li, Y.; Zhu, C.; Li, Q. Analysis of fractal dimension of coal subjected to electrical breakdown based on nuclear magnetic resonance. *J. Nat. Gas Sci. Eng.* **2020**, *79*, 103345. [CrossRef]
23. Chen, B.; Wang, H.; Hu, H.; Jiang, B. Strength model of phosphogypsum backfill based on activity rate and fractal dimension. *Ind. Miner. Process.* **2021**, *06*, 1–10.
24. Jin, S.; Zhang, J.; Han, S. Fractal analysis of relation between strength and pore structure of hardened mortar. *Constr. Build. Mater.* **2017**, *135*, 1–7. [CrossRef]
25. Deo, O.; Neithalath, N. Compressive Behavior of Pervious Concretes and a Quantification of the Influence of Random Pore Structure Features. *Mater. Sci. Eng. A* **2010**, *528*, 402–412. [CrossRef]
26. Changjiang Academy of Sciences. *Yangtze River Water Conservancy Commission, Rock Test. Rules for Water Conservancy and Hydropower Projects*; Changjiang Academy of Sciences: Wuhan, China, 2001.
27. China Institute of Water Resources and Hydropower Research. *Nanjing Institute of Water Resources Science, Hydraulic Concrete Test. Rules*; China Institute of Water Resources and Hydropower Research: Beijing, China, 2006.
28. Li, J.; Zhou, K.; Zhang, Y.; Xu, Y. Experimental study on freeze-thaw damage of rock pore structure based on nuclear magnetic resonance technique. *Chin. J. Rock Mech. Eng.* **2012**, *31*, 1208–1214.
29. Li, J.; Zhu, L.; Zhou, K.; Liu, H.; Cao, S. The damage characteristics of sandstone pore structure under freeze-thaw action. *Rock Soil Mech.* **2019**, *40*, 3524–3532.
30. She, A. *Evolution of Water State in Cement Slurry and the Relationship between Water Process and Microstructure of Slurr*; Tongji University: Shanghai, China, 2011.
31. Zhongwei, W.U. *High Performance Concrete*; China Railway Publishing House: Beijing, China, 1999; p. 24.
32. Sun, W.; Zuo, Y.; Wu, Z.; Liu, H.; Xi, S.; Shui, Y.; Wang, J.; Liu, R.; Lin, J. Fractal analysis of pores and the pore structure of the Lower Cambrian Niutitang shale in northern Guizhou province: Investigations using NMR, SEM and image analyses. *Mar. Pet. Geol.* **2018**, *99*, 416–428. [CrossRef]
33. Zhao, Y.; Zhu, G.; Dong, Y.; Danesh, N.N.; Chen, Z.; Zhang, T. Comparison of low-field NMR and microfocus X-ray computed tomography in fractal characterization of pores in artificial cores. *Fuel* **2017**, *210*, 217–226. [CrossRef]
34. Wu, R. *Performance and Mechanism Analysis of Quartz-Type Iron Tailings Powder and Waste Rock in Cement-Based Materials*; University of Science and Technology Beijing: Beijing, China, 2020.
35. Liu, J.; Zhang, Y.; Liu, R.; Duan, C. Effect of fly ash on the structure and strength of concrete. *ICEEP* **2017**, *143*, 56–63.

minerals

MDPI

Article

Application of Lignite Combustion Waste Slag Generated in Heating Plants as a Partial Replacement for Cement. Part II: Physical–Mechanical and Physical–Chemical Characterization of Mortar and Concrete

Milan Kragović [1,*], Nenad Ristić [2], Jelena Gulicovski [1], Andrijana Nedeljković [1], Snežana Pašalić [1], Ivica Ristović [3] and Marija Stojmenović [1]

1 "Vinča" Institute of Nuclear Sciences, National Institute of the Republic of Serbia, University of Belgrade, 11351 Belgrade, Serbia; rocenj@vinca.rs (J.G.); andrijana.nedeljkovic@vinca.rs (A.N.); snezana.pasalic@vinca.rs (S.P.); mpusevac@vinca.rs (M.S.)
2 Faculty of Civil Engineering and Architecture, University of Niš, 18106 Niš, Serbia; nenad.ristic@gaf.ni.ac.rs
3 Faculty of Mining and Geology, University of Belgrade, 11000 Belgrade, Serbia; ivica.ristovic@rgf.bg.ac.rs
* Correspondence: m.kragovic@vinca.rs

Citation: Kragović, M.; Ristić, N.; Gulicovski, J.; Nedeljković, A.; Pašalić, S.; Ristović, I.; Stojmenović, M. Application of Lignite Combustion Waste Slag Generated in Heating Plants as a Partial Replacement for Cement. Part II: Physical–Mechanical and Physical–Chemical Characterization of Mortar and Concrete. *Minerals* **2021**, *11*, 925. https://doi.org/10.3390/min11090925

Academic Editors: Carlito Tabelin, Chongchong Qi, Guichen Li, Hakan Basarir, Qiusong Chen and Yuantian Sun

Received: 17 June 2021
Accepted: 24 August 2021
Published: 27 August 2021

Abstract: The presented study is a continuation of the research with the aim of finding a useful value of hazardous waste slag generated by the combustion of lignite in heating plants and its application in the construction industry. The different amounts of cement (10%, 15%, 20% and 25%) were replaced with waste slag and silica fumes in mortars and concrete production. Detailed physical–mechanical characterization was performed on the mortar and concrete samples according to standard procedures. Test results indicated that the replacement of cement with slag and silica fumes reduces the physical and mechanical properties of mortar and concrete, but cement composites retained the required structural properties. If 15–20% is considered an acceptable level of compressive strength decrease, then it can be concluded that waste slag can be implemented in practice and be used as a construction material, with cement replacement in the maximal amount of 20% (17.8% of slag and 2.2% of silica fumes). On hardened mortar samples with maximal possible cement replacement (20%), physical–chemical characterizations were performed and included X-ray and infrared spectrophotometry, scanning electron microscopy, and thermal analysis. Results showed the absence of new phases and the presence of only those which were characteristic for starting samples, predominantly portlandite, quartz, calcite and calcium silicate-oxide.

Keywords: lignite combustion; waste slag; cement replacement; mortar and concrete production; physical–mechanical and physical–chemical characterization

1. Introduction

Coal is a solid fuel that has been used as a primary fossil fuel for centuries. The world's known coal reserves are much greater than those of gas and oil which means that it will be dominant in energy supply in the subsequent years [1–3]. Depending on the time of formation, coal can be classified as anthracite, bituminous coal, and lignite [4]. Lignite is a type of coal from the early stages of coalification whose properties are intermediate between peat and bituminous coal. Because of its low mining cost, easy access, high reactivity, and low price, the world's annual lignite production and consumption is approximately 1,000,000,000 tons [3,5,6].

However, besides its good properties, lignite also has some significant disadvantages such as calorific value, high-combustion waste content, the release of polluting gases (SO_x, CO_x, P_xO_y, etc.) which may cause serious environmental problems (the greenhouse effect, soil or water contamination, etc.) [3]. The non-combustible part of lignite mainly consists of moisture and ash, of which about 85–95% is fly ash and about 10% is slag [7]. Fly ash and slag are usually rich in different heavy metals and because of this, they can be harmful for the environment and human health, and consequently may be classified as hazardous waste [8]. The high amounts of produced and consumed lignite generate large quantities of combustion (hazardous) waste worldwide. Such significant quantities of hazardous waste undoubtedly have a vast negative impact on the environment, and because of that, it is particularly important to find a way for its safe storage and disposal. Finding the useful value of such generated waste as cement replacement in the construction industry is certainly a step in the right direction.

However, concerning the application of waste in the construction industry, previous research has largely been focused on finding the useful value of fly ash [9], blast furnace slag of different origins [10,11], or slag and ash from the metallurgical industry [12,13]. Additionally, there are some studies which described that the tailings materials generated in gold or copper mines may be used as construction materials, whereby the best results can be obtained after a tailings stabilization via suitable geo-polymerization [14,15]. At the same time, to our knowledge, the slag generated as waste in municipal heating plants due to lignite combustion, has not yet been sufficiently tested. This slag is very poorly described in the literature, and it is very difficult to find results related to finding its potential useful value. Additionally, it is important to emphasize the fact that for waste materials, constant research must be carried out. Very often, waste generated from different raw materials, origins (deposits, or lots), or treatments can have completely contrasting properties. For these reasons, each landfill must be monitored separately, and complete research must be carried out in order to find the potential useful value of such waste.

Accordingly, in order to solve environmental problems related to hazardous waste slag from lignite combustion, the potential use of waste slag generated in the heating plant in Valjevo, Serbia, as a replacement for cement in the production of mortar and concrete in the construction industry, was examined. Firstly, the detailed physical–chemical and physical–mechanical characterizations of slag were performed [16]. In this paper, Nedeljković et al. [16] clearly showed the properties of waste slag and gave explanations and suggestions on how waste slag, which in raw form did not meet the prescribed conditions in terms of its characteristics, could be used as a replacement for cement. Namely, the results showed that the lack of slag in terms of the content of SiO_2, Al_2O_3, Fe_2O_3, and unburned carbon, and the loss of ignition, may be overcome by the thermal treatment of the slag prior to use, up to 650 °C. Results also showed that other physical–chemical characteristics of slag (crystal, structural, and textural properties) completely satisfied the criteria prescribed by the standard EN 450-1:2012. The physical–mechanical testing showed that slag does not possess pozzolanic activity, and this was solved by mixing the slag with silica fumess in minimal amounts (slag/silica fumes = 88%/12%). The maximal possible cement replacement was 20% with 17.8% of slag and 2.2% of silica fumes. The leaching tests were also performed in accordance with the standard EN 12457-2 on the mortar with 20% of cement replacement. The leaching was carried out on crushed hardened mortar with water by mixing the mixture for 24 h, with the liquid/solid ratio of 10 dm^3/kg. The results showed the negligible release of heavy metals whose values were much lower in comparison with values prescribed by the "Rulebook on permitted quantities of hazardous and harmful materials in agricultural land and water for irrigation and methods for testing", applicable in Serbia which meant that the obtained mortar met all the required health and environmental criteria.

This paper is a continuation of that research. After the characterization of waste slag and finding ways for its potential application in the construction industry, it was important to investigate mortars and concretes in which the partial replacement of cement with waste slag was carried out. For that purpose, the optimal conditions for the practical application in mortar and concrete production are defined. Tests related to the use of waste slag as a partial replacement for cement in the production of mortars and concrete are performed in accordance with certain standards, and the detailed physical–mechanical and physical–chemical characterization of mortar and concrete is carried out.

2. Materials and Methods

2.1. Sampling

The waste slag generated by the combustion of lignite was sampled from the landfill of the heating plant located in Valjevo, Serbia. The sampling was performed as it is described elsewhere [16]. Since in the plant there is no planned waste disposal, in order to get a more realistic picture in terms of the quality of waste material, samples were taken by the square method from different places so that material of different ages was included. After sampling, the samples were mixed, homogenized and the required amount for analysis was taken by quartering.

Portland cement (CEM I 42.5R) was bought from the company CRH Popovac, Serbia. Silica fumes were bought from the company "Sika". The physical–chemical, chemical and mineralogical composition of the used cement, slag and silica fumes are given in Table 1.

Table 1. The physical–chemical, chemical and mineralogical properties of the used cement, slag and silica fumes [16].

Properties/Content	Cement	Slag	Silica Fume
ρ, g/cm^3	3.15	2.24	2.21
S_{BET}, m^2/g	1.22	25.00	28.14
SiO_2, %	21.62	21.20	93.60
Fe_2O_3, %	2.60	10.81	0.21
Al_2O_3, %	7.00	15.70	0.27
CaO, %	60.16	6.32	0.05
MgO, %	2.34	2.50	0.05
SO_3, %	2.55	1.78	0.80
Na_2O, %	0.33	2.67	0.23
K_2O, %	0.66	0.63	0.50
P_2O_5, %	-	0.036	-
LOI, %	2.68	37.45	2.40
Mineral composition	brownmillerite, calcium silicate oxide, calcite, larnite, magnesium silicate, calcium hydroxide	quartz, olivine, magnetite, pyrite, tridymite, feldspars	quartz

For making mortar, fine river sand (fraction 0/2 mm) was used. Specific gravity and water absorption of the used sand were 2610 kg/m^3 and 1.52%, respectively. For making concrete, a crushed gabbro was used as the coarse aggregates (fractions 4/8 mm and 8/16 mm) which had a specific gravity of 2890 kg/m^3 and a water absorption of 0.58%. River sand with a nominal maximum size of 4 mm (fraction 0/4 mm), the specific gravity of 2620 kg/m^3, and water absorption of 1.42% was used as the fine aggregates.

2.2. Concrete and Mortar Preparation

Preparation of mortar mixtures and production and curing of specimens for testing the physical–mechanical characteristics of hardened mortar was performed according to the standard EN 191-1. A total of 5 different mortar mixes were made: reference mixtures (E-M) and four mixtures (PM-10, PM 15, PM 20, and PM-25) in which cement was replaced with 10%, 15%, 20%, and 25% of slag and silica fumes regarding mass (whereby slag participated with 88%, and silica fumes with 12%). The mass ratio between the quantity of binder and sand in all mixtures was 1:3, while the mass ratio between the water and binder was 1:2. Superplasticizer was used to improve the workability of the mortar mixtures, and the same amount was added to all mixtures. Table 2 shows quantities of material for making one series of mortar specimens (3 prisms with dimensions 40 mm × 40 mm × 160 mm) in the laboratory mixer "Hobart N-50".

Table 2. Mix proportion of mortar for one series of samples.

Component	E-M	P10-M	P15-M	P20-M	P25-M
Cement [g (cm^3)]	450.00 (142.86)	402.34 (127.73)	378.73 (120.23)	355.27 (112.78)	331.96 (105.38)
Slag [g (cm^3)]	-	39.79 (17.76)	59.48 (26.55)	79.05 (35.29)	98.48 (43.96)
Silica fumes [g (cm^3)]	-	4.90 (2.23)	7.36 (3.33)	9.77 (4.42)	12.18 (5.51)
River sand 0/2 mm [g (cm^3)]	1350.00 (517.24)	1341.14 (513.85)	1336.70 (512.15)	1332.26 (510.44)	1327.83 (508.75)
Water [g]	225.00 (225.00)	223.52 (223.52)	222.78 (222.78)	222.04 (222.04)	221.30 (221.30)
Superplasticizer [g (cm^3)]	3.60 (3.30)	3.58 (3.28)	3.56 (3.27)	3.55 (3.26)	3.54 (3.25)
SUM [g (cm^3)]	2028.60 (888.40)	2015.29 (888.37)	2008.62 (888.24)	2001.94 (888.24)	1995.28 (888.16)

Preparation of concrete mixtures, production and curing of specimens for testing of compressive strength of concrete was performed according to the standard EN 12390-2. A total of 5 different mixtures were used, in which cement was replaced with slag and silica fumes in the same mass percentage as was the case in making the mortar mixtures. All mixtures were made with the same water–binder ratio w/b = 0.486 and the same mass ratio of aggregate and binder m_{agg}/m_b = 5.003. Superplasticizer was used for the improvement of the consistency of fresh concrete and the same quantity was added to all concrete mixtures. Mix proportions of concrete mixtures for 1 m^3 of concrete are given in Table 3.

Table 3. Mix proportion of concrete mixtures for 1 m^3 of concrete.

Component	E-C	P10-C	P15-C	P20-C	P25-C
Cement [kg (m^3)]	375.00 (0.119)	336.32 (0.107)	316.78 (0.101)	297.32 (0.094)	277.67 (0.088)
Slag [kg]	-	33.16 (0.015)	49.65 (0.022)	66.08 (0.029)	82.34 (0.037)
Silica fumes [kg (m^3)]	-	4.15 (0.002)	6.21 (0.003)	8.26 (0.004)	10.30 (0.005)
River sand 0/4 mm [kg (m^3)]	752.00 (0.287)	748.26 (0.286)	746.77 (0.285)	745.29 (0.284)	743.08 (0.284)
Crushed aggr. (4/8 mm) [kg (m^3)]	432.00 (0.149)	429.85 (0.149)	429.00 (0.148)	428.15 (0.148)	426.88 (0.148)
Crushed aggr. (8/16 mm) [kg (m^3)]	692.00 (0.239)	688.56 (0.238)	687.19 (0.238)	685.83 (0.237)	683.79 (0.237)
Water [kg (m^3)]	182.25 (0.182)	181.34 (0.181)	180.98 (0.181)	180.62 (0.181)	180.09 (0.180)
Superplasticizer [kg (m^3)]	3.75 (0.003)	3.73 (0.003)	3.72 (0.003)	3.72 (0.003)	3.71 (0.003)
SUM	2437 (0.981)	2425.37 (0.981)	2420.31 (0.981)	2415.26 (0.981)	2407.86 (0.981)
Assumed air content	~2%	~2%	~2%	~2%	~2%
Component	**E-C**	**P10-C**	**P15-C**	**P20-C**	**P25-C**
Cement [kg (m^3)]	375.00 (0.119)	336.32 (0.107)	316.78 (0.101)	297.32 (0.094)	277.67 (0.088)
Slag [kg]	-	33.16 (0.015)	49.65 (0.022)	66.08 (0.029)	82.34 (0.037)
Silica fumes [kg (m^3)]	-	4.15 (0.002)	6.21 (0.003)	8.26 (0.004)	10.30 (0.005)
River sand 0/4 mm [kg (m^3)]	752.00 (0.287)	748.26 (0.286)	746.77 (0.285)	745.29 (0.284)	743.08 (0.284)
Crushed aggr. (4/8 mm) [kg (m^3)]	432.00 (0.149)	429.85 (0.149)	429.00 (0.148)	428.15 (0.148)	426.88 (0.148)
Crushed aggr. (8/16 mm) [kg (m^3)]	692.00 (0.239)	688.56 (0.238)	687.19 (0.238)	685.83 (0.237)	683.79 (0.237)
Water [kg (m^3)]	182.25 (0.182)	181.34 (0.181)	180.98 (0.181)	180.62 (0.181)	180.09 (0.180)
Superplasticizer [kg (m^3)]	3.75 (0.003)	3.73 (0.003)	3.72 (0.003)	3.72 (0.003)	3.71 (0.003)
SUM	2437 (0.981)	2425.37 (0.981)	2420.31 (0.981)	2415.26 (0.981)	2407.86 (0.981)
Assumed aer content	~2%	~2%	~2%	~2%	~2%

2.3. Characterization of Mortar and Concrete Samples

2.3.1. Tests on Mortar Samples

Consistency was determined by the flow table in accordance with the standard EN 1015-3. For the testing, a metal conical mold was used, with a lower base diameter of 100 mm. The mean value of the mortar slump in two orthogonal directions on the flow table represented the test result.

The bulk density of fresh mortar was determined according to standard EN 1015-3, in a 1 dm^3 cylindrical metal vessel with 125 mm diameter.

The bulk density of hardened mortar was determined in accordance with standard EN 1015-10, on the specimens used for testing of mechanical strengths, e.g., on the prisms with dimensions 40 mm × 40 mm × 160 mm.

Flexural strength was determined in accordance with standards EN 196-1, EN 1015-11 and was tested on the prisms with dimensions 40 mm × 40 mm × 160 mm. The tests were conducted on the hydraulic press UTCM-6710 (Utest, Ankara, Turkey) with a measuring range of 0–10 kN of the flexural force and accuracy class 0.5%. Compressive strength was determined according to standards EN 196-1 and EN 1015-11 and was tested on the halves of the prisms obtained after testing flexural strength. The test was conducted on the hydraulic press UTEST UTCM-6710 with a measuring range of 0 to 250 kN of the compressive force and accuracy class 0.5%. Water absorption (determination of water absorption at atmospheric pressure) was determined according to EN 13755 and was conducted on the prisms with dimensions 40 mm × 40 mm × 160 mm. Prisms were firstly dried at the temperature of 70 ± 5 °C to the constant mass, and then were saturated in water. The percentage difference of masses of a saturated and a dry sample, and the mass of the dry sample represents the water absorption at atmospheric pressure.

Water absorption (determination of water absorption coefficient due to capillary action of hardened mortar) was performed according to standard EN 1015-18 and was conducted on the halves of the prisms having dimensions 40 mm × 40 mm × 160 mm. Prisms were firstly dried at a temperature of 60 ± 5 °C to the constant mass (M_0). Then the surface of the prism failure was set on the linear supports in a deep vessel, where water was poured so that the prism was immersed by 5 to 10 mm. Prior to the immersion, the mass of the dry sample was measured, and after the immersion, the mass of the sample was measured after 10 min (M_1), 90 min (M_2) and 24 h (M_3). Water absorption coefficient due to capillary action was calculated according to the following formulas:

$$\text{for coating mortars}: \ C = 0.1(M_2 - M_1) \left[kg/m^2 min^{-0.5}\right] \quad (1)$$

$$\text{for repairing mortars}: \ C = 0.625(M_3 - M_0) \left[kg/m^2\right] \quad (2)$$

Drying shrinkage was determined in accordance with standard SRPS B.C8.029:1979 (ASTM C 596) on mortar prisms and was performed on the samples having dimensions 40 mm × 40 mm × 160 mm with reference marks at their ends allowing for the monitoring of variation in length. After making mortar, mortar prisms were cured for 24 h in a mold in a water bath, after which they are removed from the mold and cured in water for the next 48 h. Baseline reading was conducted 72 h after making the prisms, and then the readings were performed after 4, 7, 14, 21 and 28 days. The difference in reading represented a variation in the length of the tested sample, while the shrinkage of cement mortar was expressed as a ratio of variation in length and the initial length of the specimen at the start of testing. The difference in readings represented the variation in length of the tested sample, while shrinkage of the cement mortar was expressed as the ratio of length variation and the initial length of the sample at the start of testing. Mortar shrinkage $\varepsilon_{sm,i}$ was calculated according to the following equation:

$$\varepsilon_{sm,i} = \frac{\Delta l_{sm}(t)}{l_{sm}} [mm/m] \quad (3)$$

where Δl_{sm} (t) was the difference of the sample length at the moment that t and the baseline reading t = 72 h in mm; l_{sm} was the length of the mortar prism expressed in m; and $\varepsilon_{sm,i}$ was the shrinkage of the mortar prism expressed in mm/m.

Adhesion (determination of adhesive strength of hardened rendering and plastering mortars on substrates) was determined according to standard EN 1015-12. A layer of mortar 10 ± 1 mm was applied to the concrete substrate. Mortar and concrete surfaces were cut to 2 mm of height using the cylindrical drill having 50 mm diameter. Afterward, a steel dolly 50 mm in diameter was glued to the mortar using epoxy glue and the dolly was pulled off using the pull-off tester. The pull-off failure force was registered, as well as the location where the failure occurred. The adhesive strength of mortar to a concrete substrate was calculated according to the following formula:

$$f_u = \frac{F}{A} \tag{4}$$

where F was the failure force at pull-off and A was the surface area of the dolly.

2.3.2. Tests of Physical–mechanical Properties of Concrete Samples

Consistency–slump test was performed in accordance with standard EN 12350-2 and was determined using the Abrams cone slump method.

The density of fresh concrete was determined according to standard EN 12350-6, in a metal, cylindrical vessel with a capacity of 8 dm^3.

Air content in fresh concrete was determined according to standard EN 12350-7 by using the pressure method. For testing, an apparatus was used which consisted of a metal cylindrical vessel with a capacity of 8 dm^3 and a special lid with a pump and manometer.

The density of hardened concrete (water-saturated) was determined according to standard EN 12390-7 on the samples shaped like cubes with a side length of 150 mm.

Flexural strength was determined in accordance with standard EN 12390-5 and testing was conducted on the prisms 400 mm in length, with a square cross-section sides of 100 mm. A digital hydraulic press "UTC-5600" (Utest, Ankara, Turkey) was used for the testing. A special tool was used for the testing permitting the loading of the sample by a bending force at two thirds of the span. Compressive strength was determined according to standard EN 12390-3 on the samples shaped like cubes with a side length of 150 mm. For the testing, digital, hydraulic press "UTC-5740" (Utest, Ankara, Turkey) with measuring range 0–3000 kN and an accuracy class of 0.5 % was used. The press was powered by servo-controlled hydraulic controller "UTC-4860" (Utest, Ankara, Turkey) and appropriate computer software. The test was conducted at the concrete age of 2, 7, 28, and 90 days.

2.3.3. Physical–chemical Characterization of Hardened Mortar Samples

The physical–chemical characterization was performed on hardened mortar samples with maximal possible cement replacement (20%) (P-20M) and included X-Ray-Structural Diffraction Analysis (XRD), infrared spectroscopy (FTIR), field emission scanning electron microscopic analysis (FESEM), and thermal analysis.

The structural properties and the composition of the crystal phase of the hardened mortar samples were determined by the XRD. For analysis, a PW-1710 X-ray diffractometer (Philips, Amsterdam, Netherlands) with a scintillation counter and curved graphite monochromator was used. The measuring was performed at room temperature at 0.02°/s, in the range of 4–65° 2θ. The voltage of 40 kV and a current of 30 mA were used, while 1° and 0.1 mm slits were used to direct the primary and diffracted beams.

The Thermo Scientific "Nicolet iS50" instrument (Thermo Fisher, Waltham, MA, USA) was used for infrared spectroscopic analyses (FTIR) of hardened mortar samples. Investigations were performed in the transmission mode. As background, the spectrum of the KBr was used. The wavelength range was 400–4000 cm^{-1}, while the resolution and number of scans were 2 cm^{-1} and 64, respectively. Two corrections: baseline automatic correction and suppression of the atmosphere were used after recording the spectra.

The field emission scanning electron microscopy and instrument TESCAN Mira3 XMU at 20 kV (Tescan, Brno–Kohoutovice, Czech Republic) was used for the investigation of the morphological and microstructural properties of the hardened mortar sample prepared under the procedure described in Table 2. Prior to analysis, the samples were pre-coated with a several nanometers-thick gold layer, using sputter coater Polaron SC503 Fision Instrument (Quorum Technologies Ltd., Lewes, UK).

Simultaneously, non-isothermal thermogravimetric analysis (TGA) and differential thermal analysis (DTA) of the hardened mortar samples was performed on a Setaram

Setsys Evolution 1750 instrument (Setaram, Lion, France). Measuring was performed in the temperature range from 25 to 1000 °C with a heating rate of 10 min^{-1} in an air atmosphere, with a flow of 16 cm^3/min. The mass of the sample was 15 mg.

3. Results and Discussion

3.1. Physical–Mechanical Characterization of Mortar and Concrete

Results of the physical–mechanical characterization of samples are given in Tables 4 and 5 and Figures 1–6.

Table 4. Physical properties of fresh and hardened mortar.

Parameter	Unit	E-M	P10-M	P15-M	P20-M	P25-M
Consistency: by flow table	mm	205 ± 7.9	187 ± 7.2	169 ± 7.5	152 ± 6.4	136 ± 6.0
Bulk density of fresh mortar	kg/m^3	2248 ± 12.3	2215 ± 10.4	2186 ± 12.0	2158 ± 9.1	2126 ± 10.9
Dry bulk density of hardened mortar	kg/m^3	2176 ± 11.5	2132 ± 12.4	2073 ± 12.0	2055 ± 9.7	2031 ± 10.2
Water absorption (water absorption at atmospheric pressure)	% (m/m)	7.1 ± 0.19	7.5 ± 0.15	8.2 ± 0.17	8.1 ± 0.14	7.9 ± 0.18
Water absorption (coefficient due to capillary action of hardened mortar): for mortars other than renovation mortars	m^2 × min$^{-0.5}$	0.25 ± 0.016	0.21 ± 0.018	0.23 ± 0.014	0.23 ±0.020	0.19 ±0.015
Water absorption (coefficient due to capillary action of hardened mortar): for renovation mortars	kg/m^2	7.65 ± 0.104	5.19 ± 0.122	5.88 ± 0.134	5.67 ± 0.115	5.11 ± 0.128

Table 5. Physical properties of fresh and hardened concrete.

Parameter	Unit	E-C	P10-C	P15-C	P20-C	P25-C
Consistency: slump test	mm	240 ± 12.3	180 ± 11.6	140 ± 10.4	90 ± 8.2	20 ± 6.5
Air content in fresh concrete	%	2.5 ± 0.25	3.4 ± 0.32	4.1 ± 0.34	5.0 ± 0.33	5.6 ± 0.35
Density of fresh concrete	kg/m^3	2473 ± 17.7	2432 ± 16.2	2369 ± 15.5	2349 ± 16.0	2341 ± 14.3
Density of hardened concrete (water-saturated	kg/m^3	2465 ± 16.9	2422 ± 17.5	2360 ± 14.7	2338 ± 15.4	2331 ± 15.0

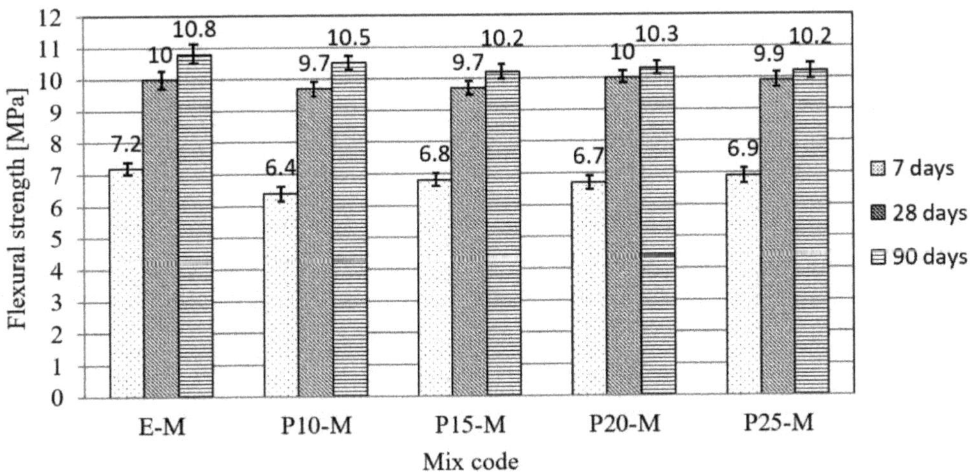

Figure 1. Flexural strength of mortar mixtures.

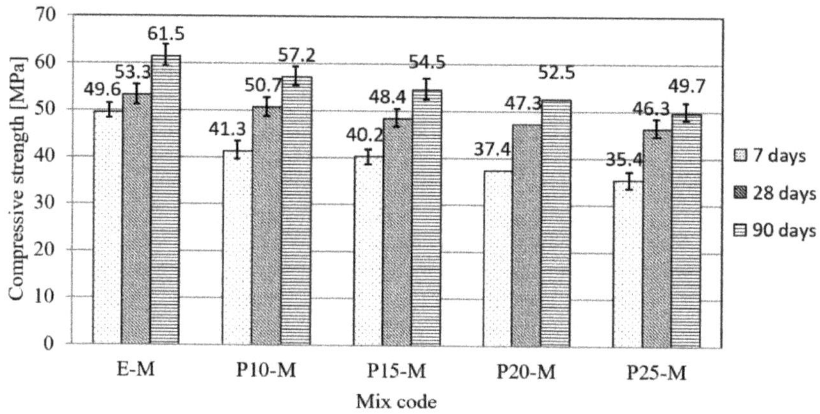

Figure 2. Compressive strength of mortar mixtures.

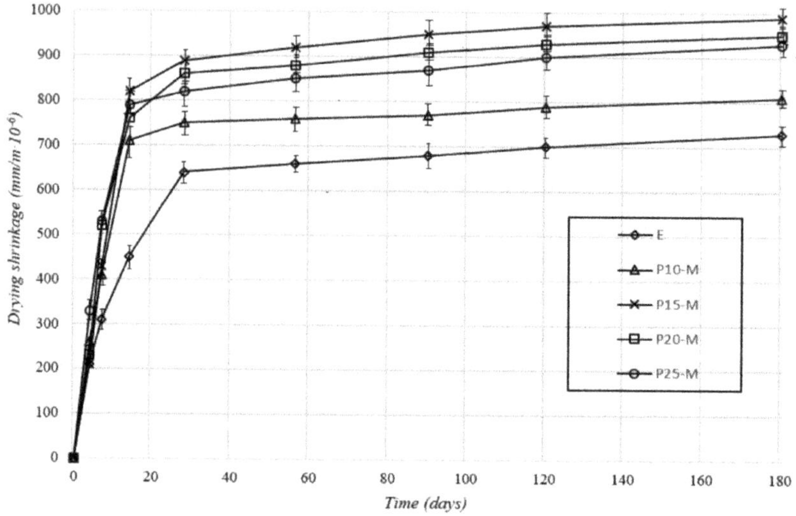

Figure 3. Drying shrinkage development of mortar mixtures.

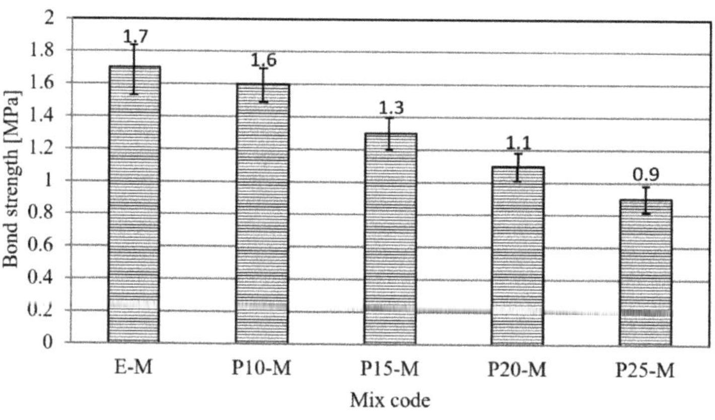

Figure 4. Adhesion between mortar and a concrete substrate.

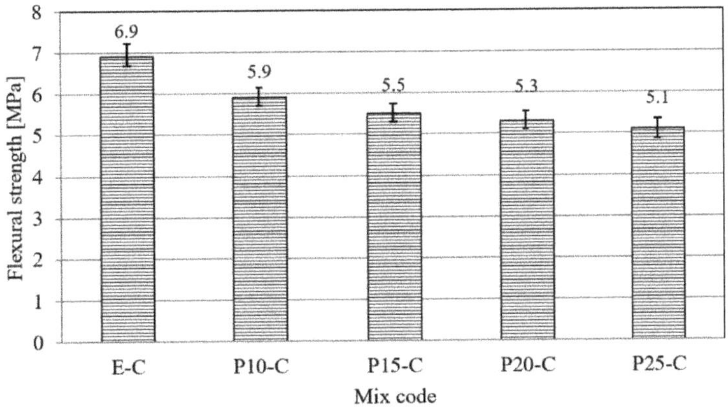

Figure 5. Flexural strength of concrete mixtures (28 days).

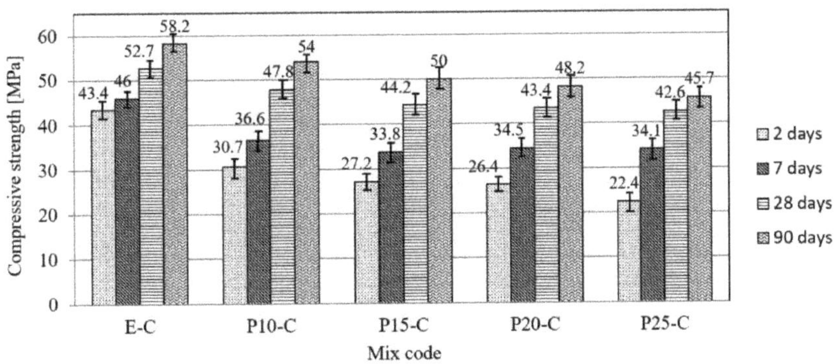

Figure 6. Compressive strength of concrete mixtures.

When testing the consistency of mortar, the slump of the mortar on a shaking table decreases with the increase in the percentage of cement replaced with slag and silica fumes. The reason is the increased water requirement of slag, which reduces the amount of free water in the mortar mixture, and thus its plasticity. By increasing the content of waste material, the bulk density of mortar in the fresh and hardened states decreases. Possible reasons for such trend were the lower specific mass of slag relative to cement, the increased porosity of the mortar due to a decrease in plasticity, as well as the lower amount of sand in mixtures with waste materials, relative to the reference mixture.

The flexural and compressive strength tests were performed at the mortar specimen age of 7, 28, and 90 days (Figures 1 and 2). Based on the test results, it can be observed that with the increasing cement replacement by waste material, the strength of mortar specimens decreased at an early age (7 days). On the other hand, with increasing age (28 and 90 days) the strength of mortar with the waste material approached the values of the reference mortar. This is attributed to the pozzolanic reaction of slag and silica fumes, which manifested itself at a later stage of hardening [17–20]. That is the reason that the percentage of possible replacement of cement with waste material was estimated based on strength at 90 days. The decrease in compressive strength was from 7.0% up to 19.2% with a percentage replacement of cement from 10% to 25%. One of the reasons for the decrease in mechanical strengths is the lower density of mortar mixtures in which cement has been replaced by waste materials.

Replacement of cement with slag resulted in an increase in the value of water absorption at atmospheric pressure, while, on the other hand, it contributed to a decrease in the absorption coefficient due to capillary water absorption. However, regardless of the test method, the maximum difference in water absorption values was up to 15%.

Based on the shrinkage measurements results due to the drying of cement mortars, (Figure 3), it can be observed that mortar mixtures with the addition of waste material had higher shrinkage values than the reference mortar, wherein the shrinkage was greater at a later curing stage. The reason for the increased shrinkage is the pozzolanic reaction of slag and silica fumes. Similar trends were observed by researchers for an examination of the influence of glass powder on the properties of cement mortar [19]. It should be mentioned that the highest shrinkage was recorded in the mortar mixture with 15% of cement replacement. This was likely because the porosities of the mortar mixtures, P20-M and P25-M, were higher than the porosity for P15-M, which means that these mixtures had a lower volume of the cement paste shrinking during hardening.

The replacement of cement with waste material had the effect of reducing the adhesion of the mortar to the concrete substrate. With an increase in the percentage of cement replacement, the adhesive strength decreased, with a decrease of 6–47% (Figure 4). The possible reason for the reduction of adhesion in mortars with a higher content of waste material is the lower quantity of free water necessary for the hydration of cement materials in the hardening period.

Based on the results of concrete consistency testing, it can be concluded that as the content of waste material increased, the size of the slump decreased significantly. This was due to the porosity and higher specific surface area (Table 1) of the waste material and its ability to retain water, thereby reducing the amount of free water in the concrete mixture, and thus the consistency [16,21,22]. The content of air entrained into fresh concrete increased with an increasing percentage of cement replacement with waste material. By increasing the content of waste material, the bulk density of concrete in the fresh and hardened states decreased, as in the case of mortar mixtures. This can be ascribed to the higher content of entrained air in fresh concrete and the lower amount of aggregate in mixtures where a partial replacement of cement was performed, and to the lower specific gravity of waste materials in comparison with cement, Table 1.

The flexural strength test was performed at the ages of samples for 28 days (Figure 5). It can be noticed that the flexural strength decreased with the increase in the replacement of cement, and that was compared to the reference concrete in the range from 14 to 26%. Compressive strength was the most significant characteristic of concrete, and it was determined at test specimen ages of 2, 7, 28, and 90 days (Figure 6). Firstly, it was observed that with the increase in cement replacement by waste material, the strengths of concrete specimens decreased at all ages of concrete. In the first 7 days, the highest increase in compressive strength was recorded with the reference mortar, while with increasing age the compressive strength of concrete prepared with waste material approached the reference concrete due to the delayed pozzolanic reaction [23]. The decrease in compressive strength at 90 days was 7.2–21.7% with a percentage replacement of cement of 10–25%. The possible reason for the reduction of the final strength of concrete with the waste material is the lower density of concrete in comparison with the reference concrete due to the higher entrained air content and the lower amount of aggregate.

Generally, the replacement of cement with slag and silica fumes reduced the physical and mechanical properties of mortar and concrete, but cement composites retained the required structural properties. Mechanical strength, and especially compressive strength, were very important parameters for cement composites. If 15–20% was considered an acceptable level of compressive strength decrease, then it can be concluded that, in mortar and concrete production, cement could be replaced by waste material by up to 20%. Additionally, results of the physical–mechanical characterization showed that mortar and concrete materials with 20% of cement replacement may be used as constructional materials in the construction industry.

3.2. X-Ray-Structural Diffraction Analysis (XRD) of the Mortar Samples

Since the results of the testing of physical–mechanical properties showed that 20% was the maximal amount of cement which may be replaced by slag and silica fumes in mortar and concrete production, physical–chemical characterization was performed on the hardened mortar sample P20-M (containing cement (17.75%), slag (3.94%), silica fumes (0.49%), river sand (66.54%), water (11.10%) and superplasticizer (0.18%)).

Results of the XRD analysis are shown in Figure 7.

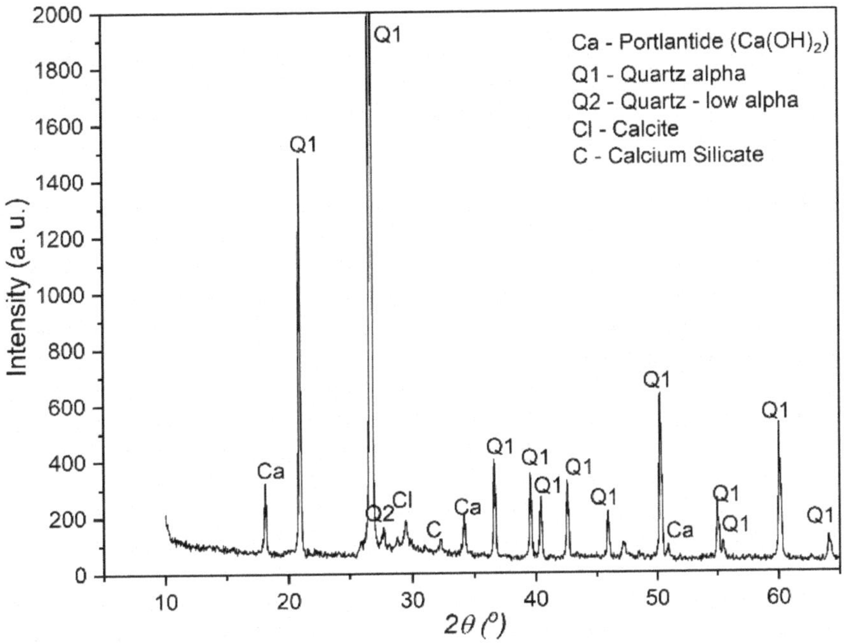

Figure 7. XRD analysis of the hardened mortar sample P20-M.

In the investigated sample the presence of the following crystal phases were observed: portlandite ($Ca(OH)_2$), quartz (SiO_2), calcite ($CaCO_3$), and calcium silicate-oxide (Ca_2SiO_4). The most abundant phase was quartz, then portlandite, and calcite, while calcium silicate-oxide was less present. Such composition was expected because portlandite, calcite and calcium silicate-oxide originated from cement, while quartz originated from river sand, silica fumes and slag. Other constituents of starting mortar components were not detected because their spectral bands overlapped with those originating from the most abundant crystalline phases. In the constituents from quartz, alpha-quartz is dominant. For hardened mortar, the crystallinity was high. Additionally, new crystalline phases except those which originated from starting samples were not detected by XRD. However, although this has not been determined in our research, the formation of new phases when waste material was used as a replacement or addition to cement was possible. For example, J. Seo et al. [24] showed that when blast furnace slag and calcium sulfoaluminate cement were mixed, a new phase strätlingite ($Ca_2Al_2SiO_7 \cdot 8H_2O$) was formed as a consequence of the reaction of aluminum–hydroxide and Ca_2SiO_4. The same phenomenon was also shown by other authors for steel slag [25] or ground perlite [26]. In order to definitively determine whether new phases had formed, additional methods of physical–chemical characterization were performed.

3.3. *Infrared Spectroscopy with Fourier Transform (FTIR)*

The infrared spectrum of the hardened mortar is given at Figure 8. In the spectrum of the hardened mortar, spectral bands at 3337, 1464, 1434, 1124, 990, 874, 850, 692, 615, 520, and 466 cm^{-1} are visible. The spectral band at 3337 cm^{-1} may be assigned to the H–H stretching vibrations due to the presence of the physical-bounded and surface-adsorbed water [27]. The bands at 1464 and 1346 cm^{-1} are characteristic of and confirmed the presence of $Ca(OH)_2$ [28]. The bands of 1434, 874, and 692 cm^{-1} are attributed to in- and out-plane bending and asymmetrical stretching vibrations of the O–C–O, respectively, which occur due to the presence of the $CaCO_3$ [29]. The bands at 1100, 797, and 615 cm^{-1} indicate the presence of the SiO_2 and are characteristic of asymmetrical and symmetrical stretching modes (1100, 797 cm^{-1}) and bending mode (615 cm^{-1}) as weak bands [30]. The spectral band at 466 cm^{-1} originates from Fe-O vibrations due to presence of iron oxide (Table 1) [31,32].

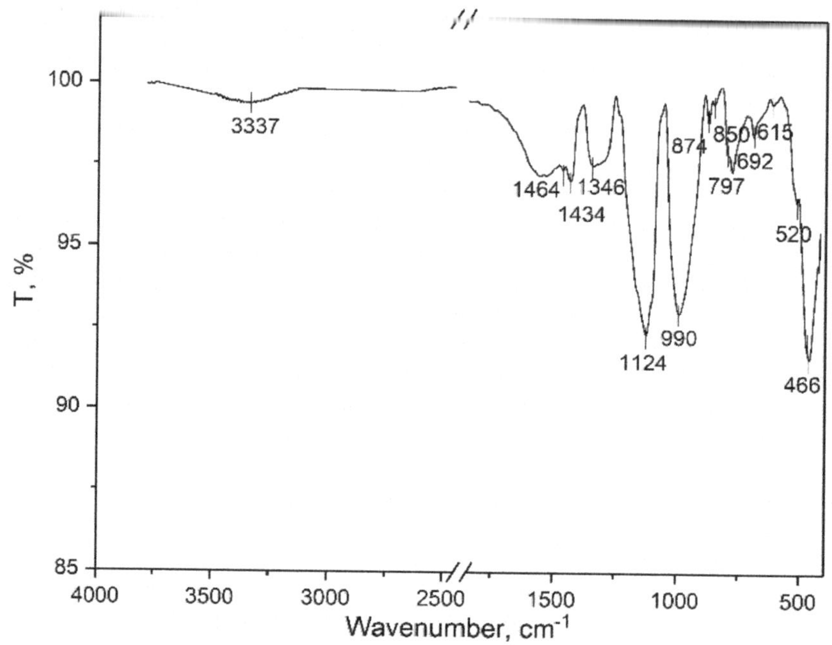

Figure 8. FTIR spectrum of the hardened mortar sample P20-M.

Other spectral bands at 990, 850, and 520 cm^{-1} are characteristic of the calcium silicate-oxide (Ca_2SiO_4) [33]. As it may be seen from the spectrum of the hardened mortar sample, only spectral bands of the main constituents detected by the XRD are visible. The spectral bands of the new functional groups, as well as bands of other constituents, are not visible which may be explained with overlapping spectral bands or what may be an indication that new phases are not formed.

3.4. *Scanning Electron Microscopy (SEM) of the Mortar Samples*

Results of the scanning electron microscopy (SEM) of the hardened mortar samples (P20-M) are given at Figure 9.

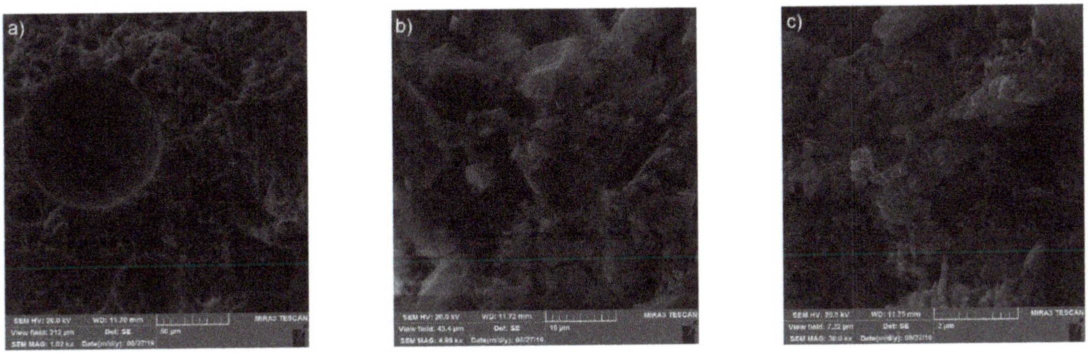

Figure 9. SEM micrographs of the hardened mortar sample P20-M taken with three different magnifications: (**a**) 1.02 k×; (**b**) 4.99 k× and (**c**) 30.0 k×.

The cementing matrix is a heterogeneous mixture of crystal phases that correspond to minerals in cement or other components in mortar. In a micrograph taken at a magnification of 1.02 k×, ball-shaped, well-defined recesses are visible which are derived from water molecules that have been ejected from the surface by the preparation and hitting of the sample prior to analysis. The micrographs taken at higher magnifications (4.99 and 30.0 k×) show a morphology in which shape and size of the platelets are characteristic of silicate, aluminosilicate and calcium hydroxide samples. The morphology shown by these investigations gives an indication that the applied design of the mortar production leads to the dispersion of cement, slag, silica fumes, river sand and superplasticizer.

3.5. Thermal Analysis of the Hardened Mortar Samples

Results of the thermal analyses (thermogravimetric and different thermal analysis (TGA-DTA)) of the designed hardened mortar sample with 20% of cement replacement are given in Figure 10.

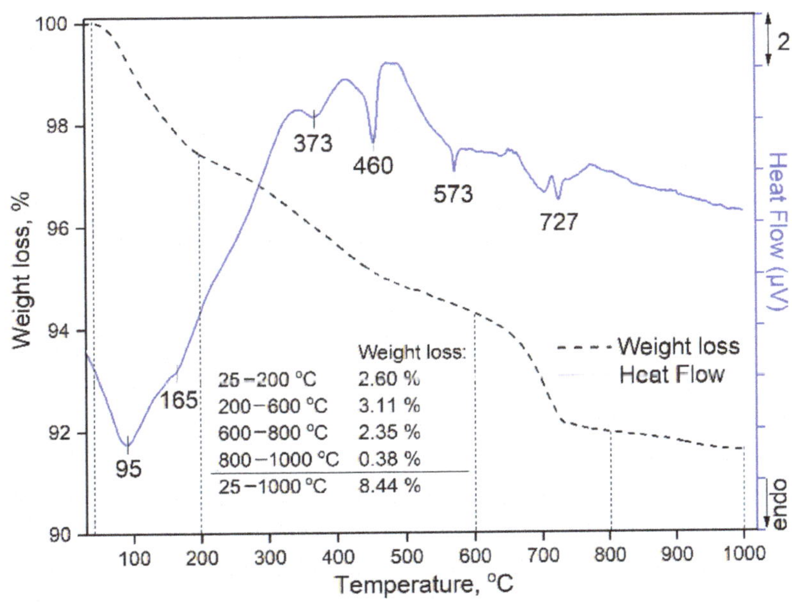

Figure 10. Thermal analysis of the hardened mortar sample P20-M. — TGA; — DTA.

From the results presented in Figure 10, it may be concluded that thermal analysis confirms the results of XRD and FTIR analysis and peaks characteristic of the phases present to the greatest extent are visible. As it may be seen from Figure 10, on the DTA curve of the investigated sample in the first temperature interval (25–200 °C), two endothermic peaks are visible, at 95 and 165 °C, of which the first one is associated with endothermic reactions such as the dehydration and removal of physically adsorbed water, while the second one originates from the dehydration and decomposition of the hydrates of the silicate-oxides, as well as calcium silicate-oxides [34]. This process has been followed with a weight loss of around 2.60% on the TGA curve. In the second temperature interval (200–600 °C) endothermic peaks are observed at 373, 460, and 573 °C. The peaks at 373 °C and 460 °C originate due to the decomposition of the calcium-hydroxide [34], while the existence of the endothermic peak at 573 °C confirms the presence of quartz (river sand) and or silicate fume (SiO_2) [35]. In this temperature interval (200–600 °C) determined mass loss by TG analysis was 3.11%. Weight loss in the third temperature interval (600–800 °C) of 2.35% and the endothermic peak at 727 °C, is caused by the decomposition of the carbonates, which is additional confirmation of the presence of the calcite ($CaCO_3$) in the designed mortar [36].

4. Conclusions

The results presented in this study show that waste slag generated in a heating plant by the combustion of lignite and taken from the Valjevo landfill in Serbia, may, despite being hazardous waste, be used in combination with silica fumes as a partial replacement for cement in the production of mortar and concrete as materials in the construction industry. Results of the physical–mechanical characterization show that maximal possible cement replacement is 20% (17.8% of slag and 2.2% of silica fumes) and for such a ratio, all requirements prescribed by the standards are satisfied.

Physical–chemical characterization is performed on the hardened mortar samples which have the maximal cement replacement (20%).

The results of the XRD analysis show that the following crystal phases are present in the investigated sample: quartz as the most abundant, then portlandite and calcite, while calcium silicate-oxide is less present.

The results of the FTIR and thermal analyses confirm the results of the XRD analysis and only the spectral bands (FTIR) and peaks (TGA/DTA) of the main constituents detected by the XRD are visible, and indicate that new phases are not formed.

The results of SEM analysis indicate that the model of mortar production applied leads to the dispersion of cement, slag, silica fumes, river sand and superplasticizer.

Author Contributions: M.K. conceived and designed the experiments, wrote the paper, and contributed to all experiments and the analyzing of the obtained results; N.R. performed physical–mechanical measurements and contributed to analyzing the results; M.S., J.G. and I.R. participated in SEM and thermal analyses and analyzed the results; A.N. and S.P. contributed in XRD measurements. All authors have read and agreed to the published version of the manuscript.

Funding: This research received no external funding.

Data Availability Statement: The study do not report any data.

Acknowledgments: This work was supported by the Ministry of Education, Science and Technological Development of the Republic of Serbia through the realization of research themes 1702103 and 1702105.

Conflicts of Interest: The authors declare no conflict of interest.

References

1. Zhang, F.; Xing, Y.; Chang, G.; Yang, Z.; Cao, Y.; Gui, X. Enhanced lignite flotation using interfacial nanobubbles based on temperature difference method. *Fuel* **2021**, *293*, 120313. [CrossRef]
2. Zhao, H.Y.; Li, Y.H.; Song, Q.; Liu, S.C.; Yan, J.; Ma, Q.X.; Zhu, X. Investigation on the thermal behavior characteristics and products composition of four pulverized coals: Its potential applications in coal cleaning. *Int. J. Hydrogen Energy* **2019**, *44*, 23620–23638. [CrossRef]

3. Xia, W.C.; Xie, G.Y.; Peng, Y.L. Recent advances in beneficiation for low rank coals. *Powder Technol.* **2015**, *277*, 206–221. [CrossRef]
4. American Geosciences Institute. Available online: https://www.americangeosciences.org/critical-issues/faq/what-are-the-dierent-types-of-coal (accessed on 30 May 2021).
5. Yu, J.L.; Tahmasebi, A.; Han, Y.N.; Yin, F.K.; Li, X.C. A review on water in low rank coals: The existence, interaction with coal structure and effects on coal utilization. *Fuel Process Technol.* **2013**, *106*, 9–20. [CrossRef]
6. Available online: https://iea.blob.core.windows.net/assets/c40f0317-f8e6-4f00-b183-27a24c7b6a8f/Coal_Information_2019_Overview.pdf (accessed on 30 May 2021).
7. Aydin, S.; Baradan, B. Efect of pumice and fly ash incorporation on high temperature resistance of cement based mortars. *Cem. Concr. Res.* **2007**, *37*, 988–995. [CrossRef]
8. European List of Waste, the Classification of Waste Based on: The European List of Waste (Commission Decision 2000/532/EC—Consolidated Version) and Annex III to Directive 2008/98/EC (Consolidated Version). Available online: http://data.europa.eu/eli/dec/2000/532/2015-06-01 (accessed on 25 May 2021).
9. Berryman, C.; Zhu, J.; Tadros, M. High-percentage replacement of cement with fly ash for reinforced concrete pipe. *Cem. Concr. Res.* **2005**, *35*, 1088–1091. [CrossRef]
10. Ray, T.; Mohanta, N.R.; Kumar, M.H.; Macharyulu, I.S.; Samantaray, S. Study of effect of temperature on behavior of alkali activated slag concrete. *Mater. Today Proc.* **2021**, *43*, 1352–1357. [CrossRef]
11. Czop, M.; Łaźniewska-Piekarczyk, B. Use of Slag from the Combustion of Solid Municipal Waste as a Partial Replacement of Cement in Mortar and Concrete. *Materials* **2020**, *13*, 1593. [CrossRef]
12. Rivera, J.F.; Cristelo, N.; Fernández-Jiménez, A.; de Gutiérrez, R.M. Synthesis of alkaline cements based on fly ash and metallurgic slag: Optimisation of the SiO_2/Al_2O_3 and Na_2O/SiO_2 molar ratios using the response surface methodology. *Constr. Build. Mater.* **2019**, *213*, 424–433. [CrossRef]
13. Zhang, Y.; Zhang, S.; Ni, W.; Yan, Q.; Gao, W.; Li, Y. Immobilisation of high-arsenic-containing tailings by using metallurgical slag-cementing materials. *Chemosphere* **2019**, *223*, 117–123. [CrossRef]
14. Tabelin, C.B.; Villacorte-Tabelin, M.; Park, I.; Phengsaart, T.; Jeon, S.; Hiroyoshi, N. Copper and critical metals production from porphyry ores and E-wastes: A review of resource availability, processing/recycling challenges, socio-environmental aspects, and sustainability issues. *Resour. Conserv. Recycl.* **2021**, *170*, 105610. [CrossRef]
15. Opiso, E.M.; Tabelin, C.B.; Maestre, C.V.; Aseniero, J.P.J.; Park, I.; Illacorte-Tabelin, M. Synthesis and characterization of coal fly ash and palm oil fuel ash modified artisanal and small-scale gold mine (ASGM) tailings based geopolymer using sugar mill lime sludge as Ca-based activator. *Heliyon* **2021**, *7*, e06654. [CrossRef]
16. Nedeljković, A.; Stojmenović, M.; Gulicovski, J.; Ristić, N.; Milićević, S.; Krstić, J.; Kragović, M. Waste Slag from Heating Plants as a Partial Replacement for Cement in Mortar and Concrete Production. Part I—Physical–Chemical and Physical–Mechanical Characterization of Slag. *Minerals* **2020**, *10*, 992.
17. Cheah, C.B.; Ramli, M. Mechanical strength, durability and drying shrinkage ofstructural mortar containing HCWA as partial replacement of cement. *Constr. Build. Mater.* **2012**, *30*, 320–329. [CrossRef]
18. Garcia, M.D.L.; Sousa-Coutinho, J. Strength and durability of cement with forest waste bottom ash. *Constr. Build. Mater.* **2013**, *2013 41*, 897–910. [CrossRef]
19. Grdić, D.; Ristić, N.; Topličić-Ćurčić, G.; Đorđević, D.; Krstić, N. Effects of addition of finely ground CRT glass on the properties of cement paste and mortar. *Građevinar* **2020**, *72*, 1–10.
20. Mafalda, A.; Sousa-Coutinho, J. Durability of mortar using waste glass powder as cement replacement. *Constr. Build. Mater* **2012**, *36*, 205–215.
21. Chavan, S.P.; Salokhe, S.A.; Nadagauda, P.A.; Patil, S.T.; Mane, K.M. An investigational study on properties of concrete produced with industrial waste red mud. *Mater. Today Proc.* **2021**, *42*, 733–738. [CrossRef]
22. Anirudh, M.; Rekha, K.S.; Venkatesh, C.; Nerella, R. Characterization of red mud based cement mortar; mechanical and microstructure studies. *Mater. Today Proc.* **2021**, *43*, 1587–1591. [CrossRef]
23. Sadiqul, I.G.M.; Rahman, M.H.; Kazi, N. Waste glass powder as partial replacement of cement for sustainable concrete practice. *Int. J. Sustain. Built Environ.* **2017**, *6*, 37–44.
24. Seo, J.; Kim, S.; Park, S.; Yoon, H.N.; Lee, H.K. Carbonation of calcium sulfoaluminate cement blended with blast furnace slag. *Cem. Concr. Compos* **2021**, *118*, 103918. [CrossRef]
25. Liao, Y.; Jiang, G.; Wang, K., Al Qunaynah, S.; Yuan, W. Effect of steel slag on the hydration and strength development of calcium sulfoalumi-nate cement. *Constr. Build. Mater.* **2020**, *265*, 120301. [CrossRef]
26. Sengül, K.; Erdogan, S.T. Influence of ground perlite on the hydration and strength develop-ment of calcium aluminate cement mortars. *Constr. Build. Mater.* **2021**, *266*, 120943. [CrossRef]
27. Mojet, B.L.; Ebbesenz, S.D.; Lefferts, L. Light at the interface: The potential of attenuated total reflection infrared spectroscopy for understanding heterogeneous catalysis in water. *Chem. Soc. Rev.* **2010**, *39*, 4643–4655. [CrossRef] [PubMed]
28. Cavallaro, G.; Danilushkina, A.A.; Evtugyn, V.G.; Lazzara, G.; Milioto, S.; Parisi, F.; Rozhina, E.V.; Fakhrullin, R.F. Halloysite Nanotubes: Controlled Access and Release by Smart Gates. *Nanomaterials* **2017**, *7*, 199. [CrossRef] [PubMed]
29. Abdolmohammadi, S.; Siyamak, S.; Ibrahim, N.A.; Yunus, W.M.Z.W.; Rahman, M.Z.A.; Azizi, S.; Fatehi, A. Enhancement of Mechanical and Thermal Properties of Polycaprolactone/Chitosan Blend by Calcium Carbonate Nanoparticles. *Int. J. Mol. Sci.* **2012**, *13*, 4508–4522. [CrossRef] [PubMed]

30. Tran, T.N.; Pham, T.V.A.; Le, M.L.P.; Nguyen, T.P.T.; Tran, V.M. Synthesis of amorphous silica and sulfonic acid functionalized silica used as reinforced phase for polymer electrolyte membrane. *Adv. Nat. Sci. Nanosci. Nanotechnol.* **2013**, *4*, 045007. [CrossRef]
31. Tabelin, C.B.; Veerawattananun, S.; Ito, M.; Hiroyoshi, N.; Igarashi, T. Pyrite oxidation in the presence of hematite and alumina: I. Batch leaching experiments and kinetic modeling calculations. *Sci. Total Environ.* **2017**, *580*, 687–698. [CrossRef]
32. Park, I.; Tabelin, C.B.; Magaribuchi, K.; Seno, K.; Ito, M.; Hiroyoshi, N. Suppression of the release of arsenic carrier-microencapsulation from arsenopyrite by using Ti-catechol complex. *J. Hazard. Mater.* **2018**, *344*, 322–332. [CrossRef]
33. Handke, M.; Ptak, W. IR and Raman studies of the stabilization beta-Ca2SiO4. *Ceramurgia international* **1978**, *4*, 75. [CrossRef]
34. Zhang, G.; Li, Z.; Zhang, L.; Shang, Y.; Wang, H. Experimental research on drying control condition with minimal effect on concrete strength. *Constr. Build. Mater* **2017**, *135*, 194–202. [CrossRef]
35. Schelz, J.P. The detection of quartz in clay minerals by differential thermal analysis. *Thermochimicu Acta* **1976**, *15*, 17-B. [CrossRef]
36. Parvan, M.-G.; Voicu, G.; Badanoiu, A.-I.; Nicoara, A.-I.; Vasile, E. CO_2 Sequestration in the Production of Portland Cement Mortars with Calcium Carbonate Additions. *Nanomaterials* **2021**, *11*, 875. [CrossRef] [PubMed]

minerals

MDPI

Article

A Method of Backfill Mining Crossing the Interchange Bridge and Application of a Ground Subsidence Prediction Model

Shuaigang Liu [1,2], Jianbiao Bai [2,*], Gongyuan Wang [1,2], Xiangyu Wang [1] and Bowen Wu [1,2]

1 School of Mines, China University of Mining and Technology, Xuzhou 221116, China; liushuaigang@cumt.edu.cn (S.L.); wanggongyuan@cumt.edu.cn (G.W.); wangxiangyu@cumt.edu.cn (X.W.); wbw_cumt@cumt.edu.cn (B.W.)
2 State Key Laboratory of Coal Resources and Safe Mining, China University of Mining and Technology, Xuzhou 221116, China
* Correspondence: baijianbiao@cumt.edu.cn

Abstract: The traditional backfill mining method is a technology developed by the general trend of green coal mining, but with a high cost and an impact on production efficiency. This paper proposes a structured backfill mining method with high-water materials and pillars. The evolution of roof pressure appearance is assessed through the sensor and monitoring system in the hydraulic support. The main roof fracture step distance is determined based on the roof structure characteristics of backfill mining, and the backfill step distance of underground structural backfill is 22.7 m considering the safety factor. Through the simulation results of Abaqus commercial simulation software, the roof subsidence evolution of different backfill schemes under temporary load and permanent load is compared, and the rationality of the backfill step distance is verified. Based on the probability integral method, the surface subsidence prediction model is proposed, then the final value and the maximum dynamic change value of the surface subsidence at the north and south ends of the interchange bridge by traditional mining and backfill mining are analyzed, which verifies the rationality of the structural backfill mining method.

Keywords: structural backfill mining; sensor and monitoring system; ground subsidence; interchange bridge

Citation: Liu, S.; Bai, J.; Wang, G.; Wang, X.; Wu, B. A Method of Backfill Mining Crossing the Interchange Bridge and Application of a Ground Subsidence Prediction Model. *Minerals* **2021**, *11*, 945. https://doi.org/10.3390/min11090945

Academic Editors: Abbas Taheri, Qiusong Chen, Chongchong Qi, Guichen Li, Hakan Basarir and Yuantian Sun

Received: 11 July 2021
Accepted: 27 August 2021
Published: 30 August 2021

1. Introduction

Ground subsidence mitigation regarding backfill mining is inextricably linked to backfill mining technology [1–5]. Towards the extent that solid backfill can be used to control ground subsidence, some scholars have conducted systematic analyses of the equivalent mining diameter theory of strata movement in backfill mining, developed a structural mechanics model of strata movement in high-density backfill mining, and performed continuous medium mechanics analysis [6–8]. It is proposed that when the probability integral approach is employed to predict ground sinking regarding dense filling, the traditional ground subsidence prediction specifications will be adjusted [9,10].

Additionally, some scholars established a model for estimating solid-filled coal mining subsidence based on the equivalent mining height theory, suggested a design procedure for solid-filled mining under buildings, as well as described a strategy for determining the probability integral method's parameters [11–13]. Vyazmensky et al. [14] and Hamdi et al. [14,15] constructed a computational model of ground subsidence in backfill mining and simulated it using discrete element and finite difference methods. Li et al. [16] developed a method for calculating subsidence regions using a surface subsidence database and described the principles, specifications, methodologies, and steps for sketching a subsidence boundary, completed a quantitative assessment of underground coal mining damage caused by surface subsidence and promoted the use of Geographic Information

System (GIS) in mining damage assessment. Shi et al. [17] constructed a simulation experiment methodology to examine the propagation of cracks in overlaying layers based on the real mechanical characteristics of rock and soil in goaf. Lamich et al. [18] proposed a novel, self-contained approach for evaluating the linear direction of potentially hazardous locations in order to estimate the impact of underground coal mining on future development. It is proposed that standard geodetic monitoring of roadways be developed as a stand-alone technique for identifying hazardous regions in foundation engineering. The laws of overburden movement, the process of overburden collapse, and the dynamic behaviors of overburden are the primary concerns of subsidence theory [15,19–24]. The laws of strata movement and surface subsidence are studied by scholars from different scopes. On the one hand, it is concerned with analyzing the mechanical mechanisms behind the behavior of mining rock masses through the examination of overburden failure, deformation, and movement laws, and then with analyzing surface deformation and failure [25–27]. On the other hand, it has created significant achievements in both theoretical and practical investigations, which concentrate on the mathematical analysis and assessment of the mining rock mass's behavior, beginning with research into the surface curvature and failure laws and progressing to the analysis of the overlaying rock's distortion and breakdown [10,23,28–30]. The strata stabilization theory is determined by the relationship between overburden deformation and failure and the ground on the ground, which differs from the relationship observed in traditional mining [18,31–33]. Hence, the strata systems theory of backfill mining provides the foundation for studying the deformation and breakdown of overburden and ground subsidence, for determining the filling method, filling material performance, and support strength, and for representing a critical comparison for cost reduction. As a result, it is essential to expand studies into the strata mitigation theory of backfill mining [25,27,30].

Currently, mining technology for "three under" coal pressure consists primarily of multiple methodologies: protective coal pillar locations, goaf backfill mining, incomplete mining, goaf gangue space grouting filling, and overburden separation zone grouting [34,35]. The traditional "three under" mining technology suffers from inefficiency and resources wastage. The Liangshuijing coal mine 421 area mined a total of 14 working faces from panel 42,101 to panel 42,114. Now, there is a challenge with "three under" coal pressure in panel 42,115. If the backfill protection mining strategy is not implemented, a considerable number of protective coal pillars will need to be installed in the 42,115 panel at the interchange bridge section. On the one hand, this will result in a significant waste of coal resources. On the other hand, when mining the 42,115 panel, the working face's equipment must be relocated, significantly reducing the working face's production efficiency. The efficient processing of 42,115 work faces while ensuring the safety and stability of the interchange bridge is a critical factor at the Liangshuijing coal mine. Thus, research on structural backfill mining methods based on roof structural characteristics is being conducted in order to give theoretical and technological guidance for ensuring interchange bridge mining protection.

Therefore, the objectives of this paper are:

- Propose a roof subsidence numerical simulation method and determine the structural backfill parameters;
- Theoretical basis and application of a ground subsidence prediction model;
- To develop an efficient structural backfill method through "three under" coal mining, to be combined with high-water materials filling barricades and pillars.

2. Geological Conditions and Structural Backfill Schemes

2.1. Geological Conditions of Working Face under Interchange Bridge

The Liangshuijing coal mine is about 16 km west of Shenmu County, Yulin City, Shaanxi Province, with the following geographic coordinates: 110°14′22″–110°21′24″ E, 38°47′29″–38°53′24″ N (shown in Figure 1a). The relationship between interchange bridge above 42,115 working face and underground mining is shown in Figure 1b, the 42,115 fully mechanized working face is located to the north of the 421 mine area. The length and width of the 42,115 panel are 4380 m and 330 m, respectively. The buried depth of the coal seam (42#) near the 42,115 and 42,114 panels is 250 m. The geological drilling near interchange bridge shows that the roof and floor rock properties of 42,115 working face are shown in Figure 2, the coal seam has a dip angle of about 1°~3° and an average thickness of about 3.5 m. The lithology is mostly mudstone and siltstone, and the coal seam structure is complex. The lithology of the roof is mainly siltstone and fine sandstone, locally medium grained sandstone, and mudstone, and occasionally mudstone pseudo roof, which is a class II medium stable roof. The lithology of the coal seam floor is mainly siltstone, locally fine sandstone, and mudstone. The average thickness of the immediate roof is about 5.0 m, and the average thickness of the main roof is about 15.0 m.

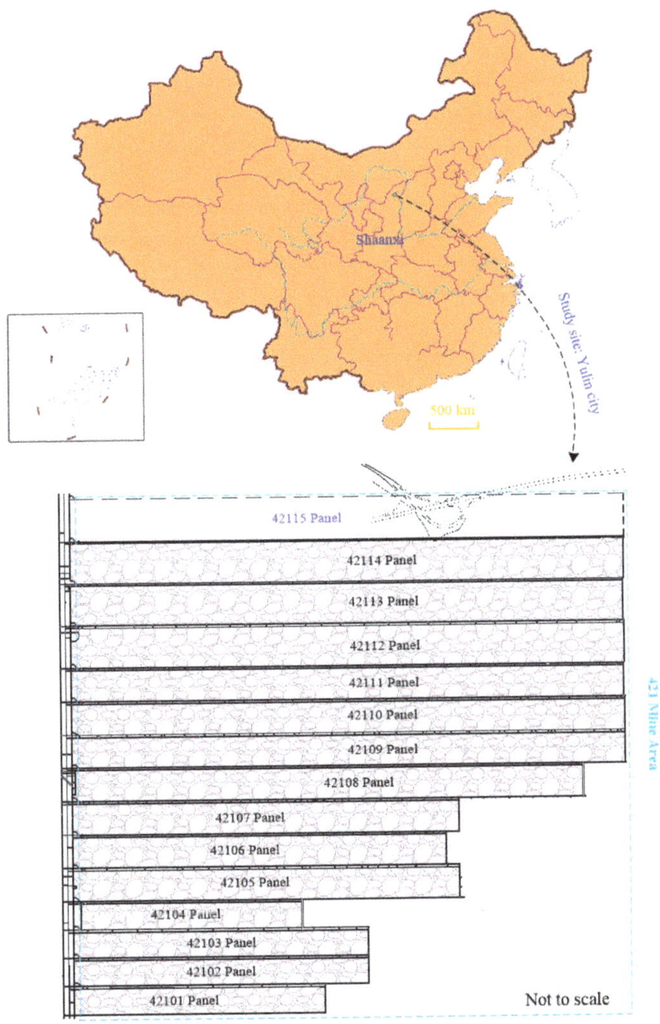

(a)

Figure 1. *Cont.*

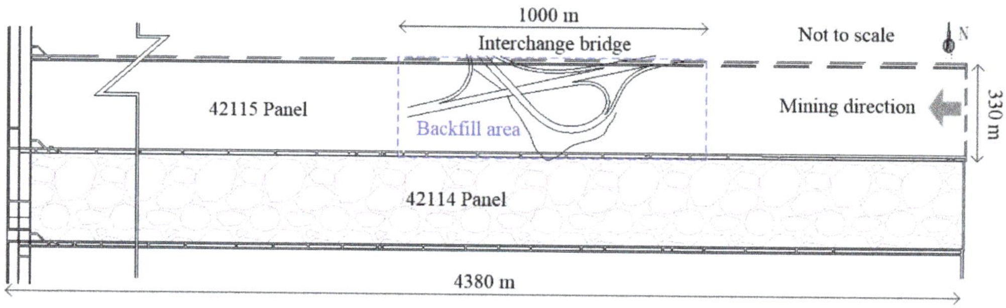

(**b**)

Figure 1. The study coal mine location and geological conditions. (**a**) Location of Liangshuijing coal mine and panels layout in the 421 mine area; (**b**) The mining conditions of 42,115 panel and adjacent panel.

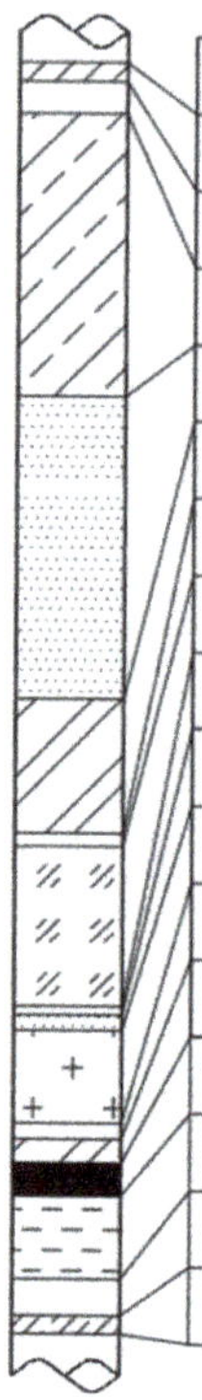

No.	Rock layer	Thickness/m	Depth/m
1	Aeolian sand	2.39	2.39
2	Silty sand	3.92	6.31
3	Loess	35.08	42.11
4	Clay	37.54	79.65
5	Fine sandstone	16.62	96.27
6	Mudstone	1.60	97.87
7	Siltstone	19.80	117.67
8	Fine sandstone	1.12	118.79
9	Siltstone	1.90	120.69
10	Fine sandstone	11.44	132.13
11	Medium sandstone	2.00	134.13
12	Fine sandstone	2.97	137.10
13	4-2 Coal seam	4.00	141.10
14	Siltstone	10.41	151.51
15	Fine sandstone	4.48	155.99
16	Medium sandstone	2.26	158.25

Figure 2. A borehole histogram is displayed in the L7 borehole.

2.2. Mine Pressure Evolution of the Working Face around the Interchange Bridge

2.2.1. Mine Pressure Evolution Monitoring and Result of the 42,114 Working Face

ZY12000/20/40D (a hydraulic support codename) hydraulic support was used in the 42,111 working face, and a German Marco control and monitoring system was installed on the support for real-time monitoring. The Marco system consists of three parts at the control level, mechatronics control at the level of single hydraulic support, fieldbus control at the level of working face and Supervisory Control and Data Acquisition (SCADA, production process control) control at the level of channel. The controllers in the fully mechanized working face are connected into a control system through the cables between the supports.

Through the bus communication technology, the fieldbus control system including all the hydraulic supports in the working face is formed. The data is transmitted to the ground monitoring host, and the monitoring results are obtained after the system program analysis and processing. The mine pressure monitoring equipment system can monitor the stope pressure behavior in real time, analyze the mine pressure data reasonably, and then forecast the failure of the roof based on the stope pressure behavior.

During the mining process of the 42,114 panel, the working resistance of the hydraulic support was monitored in the area (745 m~800 m) potentially affected by the interchange bridge, and the specific monitoring location is shown in Figure 3. The working resistance monitoring results of the upper area hydraulic support (15#, 20#), the middle area hydraulic support (70#, 78#, 115#, 121#), and the lower area hydraulic support (160#, 168#) in the 745 m~800 m monitoring area are shown in Figure 4a–h.

The evolution of hydraulic support working resistance is a common method to predict the distance of the periodic pressure of the roof [36]. Usually, the sum of the average working resistance of the hydraulic support and the variance of its average value is used as the index of the periodic pressure judgment. The judgment criteria are as follows:

$$P = \overline{P_t} + \sigma_P \tag{1}$$

$$\sigma_P = \sqrt{\frac{1}{n}\sum_{i=1}^{n}(p_{ti} - \overline{p_t})^2} \tag{2}$$

$$\overline{P_t} = \frac{1}{n}\sum_{i=1}^{n} p_{ti} \tag{3}$$

where P is the period of pressure criterion, MPa; $\overline{P_t}$ is the average value of hydraulic support resistance at the end of the cycle, MPa; σ_P is the variance of the mean value of the resistance of the hydraulic support at the end of the cycle, MPa; and p_{ti} is the working resistance of the hydraulic support at the end of the cycle, MPa.

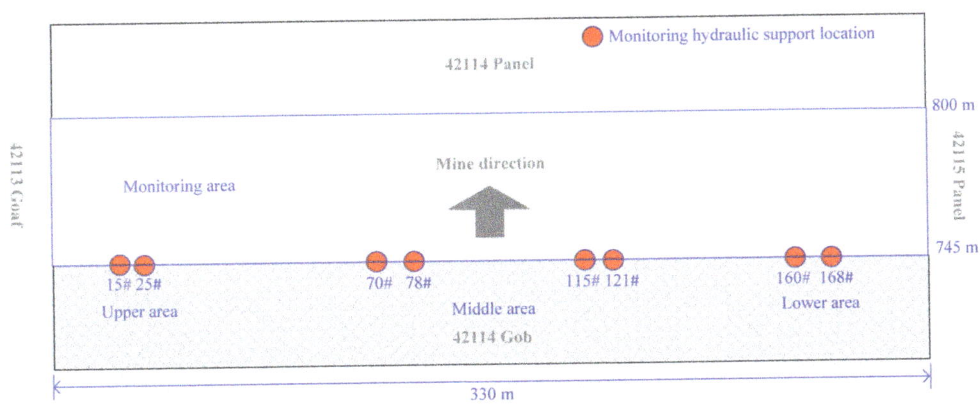

Figure 3. The monitoring location of hydraulic support.

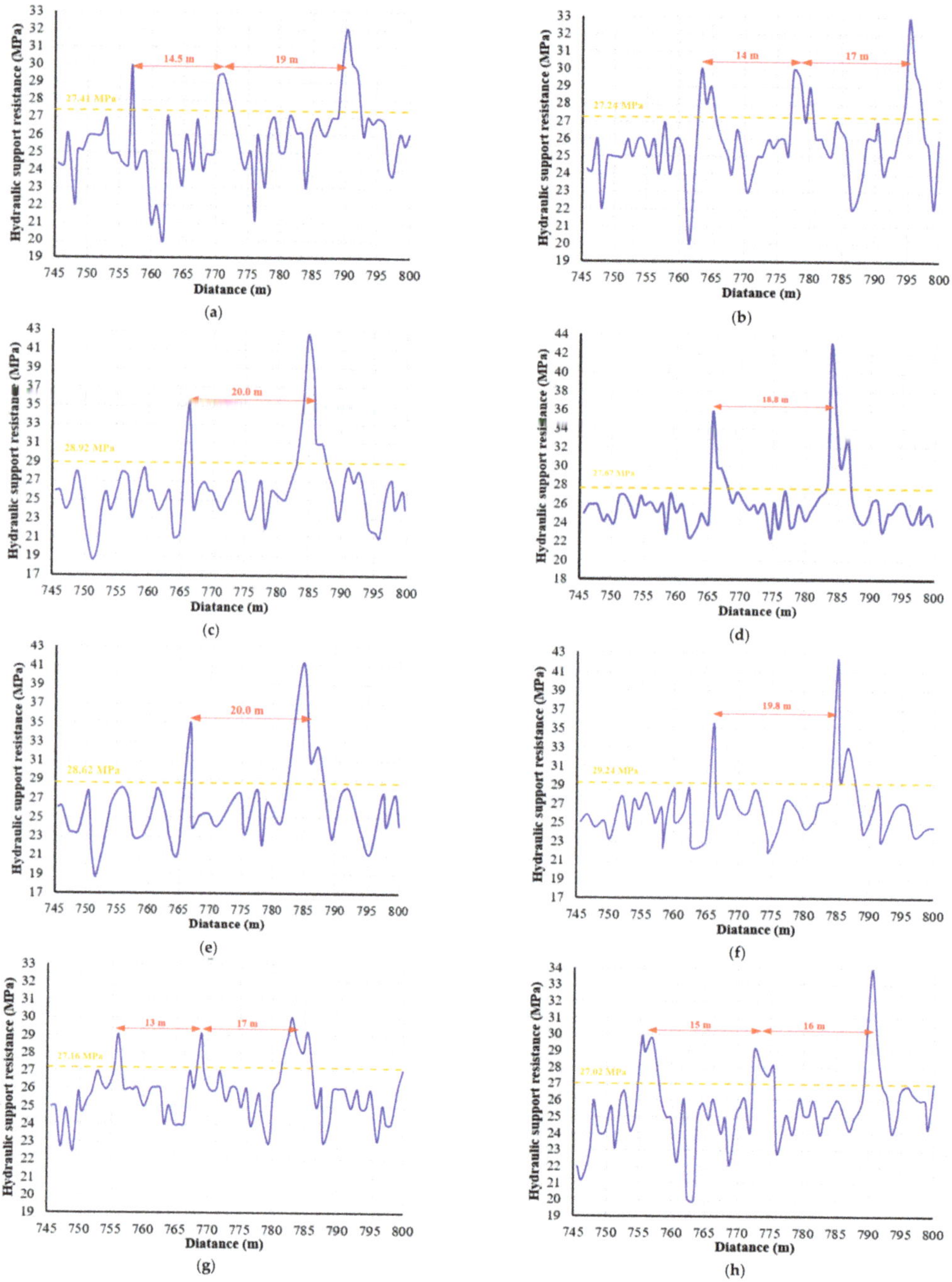

Figure 4. Monitoring results of the hydraulic support resistance during mining in 42,114 working face. (**a**) Resistance of 15# hydraulic support; (**b**) Resistance of 20# hydraulic support; (**c**) Resistance of 70# hydraulic support; (**d**) Resistance of 78# hydraulic support; (**e**) Resistance of 115# hydraulic support; (**f**) Resistance of 121# hydraulic support; (**g**) Resistance of 160# hydraulic support; and (**h**) Resistance of 168# hydraulic support.

According to Figure 4a–g, with the change of the advancing distance of the 42,114 working face, the hydraulic support resistance of the three areas shows periodic changes.

Furthermore, the upper monitoring area (Figure 4a,b) hydraulic supports maximum working resistance of 32.1 MPa, 33.0 MPa, average working resistance of 25.20 MPa, 25.14 MPa, and periodic pressure distances of 14.5 m, 19 m, and 17 m, the average periodic distance is 16.1 m; the middle monitoring area (Figure 4c–f) hydraulic supports maximum working resistance of 42.9 MPa, 43.1 MPa, 41.2 MPa, 42.5 MPa, average working resistance of 25.04 MPa, 24.45 MPa, 25.11 MPa, 24.81 MPa, and periodic pressure distances of 20 m, 18.8 m, and 19.8 m, the average periodic distance is 19.7 m; the upper monitoring area (Figure 4g,h) hydraulic supports maximum working resistance of 30.9 MPa, 33.7 MPa, average working resistance of 25.02 MPa, 24.52 MPa, and periodic pressure distances of 13 m, 17 m, 15 m, 16 m, and 14.5 m; the average periodic distance is 15.3 m. Combined with Table 1, the upper hydraulic support resistance and the lower hydraulic support resistance of the working face are slightly lower than the middle hydraulic support resistance, which is consistent with the previous research result [36,37] for the period pressure distance evolution. The periodic pressure step distance is about 15.3 m~19.7 m, with an average value of 17.0 m.

Table 1. Periodic pressure characteristic value of 42,114 working face monitoring area.

Remark	HS_n	MS_{hsec}/MPa	$\overline{P_t}$/MPa	σ_P/MPa	PPC/MPa	PPD/m
Upper area	15#	32.1	25.20	2.21	27.41	14.5, 19
	20#	33.0	25.14	2.10	27.24	14, 17
Middle area	70#	42.9	25.04	3.88	28.92	20
	78#	43.1	24.45	3.25	27.67	18.8
	115#	41.2	25.11	3.51	28.62	20
	121#	42.5	24.81	3.43	29.24	19.8
Lower area	160#	30.9	25.02	2.14	27.16	13, 17
	168#	33.7	24.52	2.50	27.02	15, 16

HS_n—Hydraulic support number; MS_{hsec}—Maximum working resistance of hydraulic support at the end of cycle; $\overline{P_t}$—Average working resistance of hydraulic support at the end of cycle; σ_P—The variance of the mean value of the resistance of the hydraulic support at the end of the cycle; PPC—Periodic pressure criterion; PPD—Periodic pressure distance.

2.2.2. General Situations of Backfill Strategies

Figure 5 depicts the relationship between the 42,115 working face and its interchange bridge. The open-off cut width of 42,115 working face is 330 m, and the mining has been completed near 42,114 working face. The width of the yielding coal pillar between the two working faces is 10 m. In order to reduce the impact of 42,115 working face mining on the stability of the interchange bridge, it is necessary to carry out certain protective mining in the corresponding area under the interchange bridge and carry out structural backfill in a certain direction of the 42,115 working face. Under the protection of structural backfill, it is necessary to ensure that 42,115 working face mining has as little impact on the interchange bridge as possible. The feasible backfill technology scheme of 42,115 working face passing through the interchange bridge has also been determined.

When the 42,115 working face is mined to the backfill position, the roof is maintained by means of net laying, roof rock bolts, and anchor cables. The channel formed by the backfill support is used to transport the single hydraulic prop and wooden stack to the goaf. At the same time, the high-water materials are used to fill the isolation wall, so that the height of the roof and floor can reach 3.4 m. To fill the steel pipes, new high-water materials with high strength and pumpability are used. The cast-in-place support pillars and the gangue (high-water materials filling) between the pillars are used to support the roof. The initial roof support parameters in the backfill area are shown in Table 2.

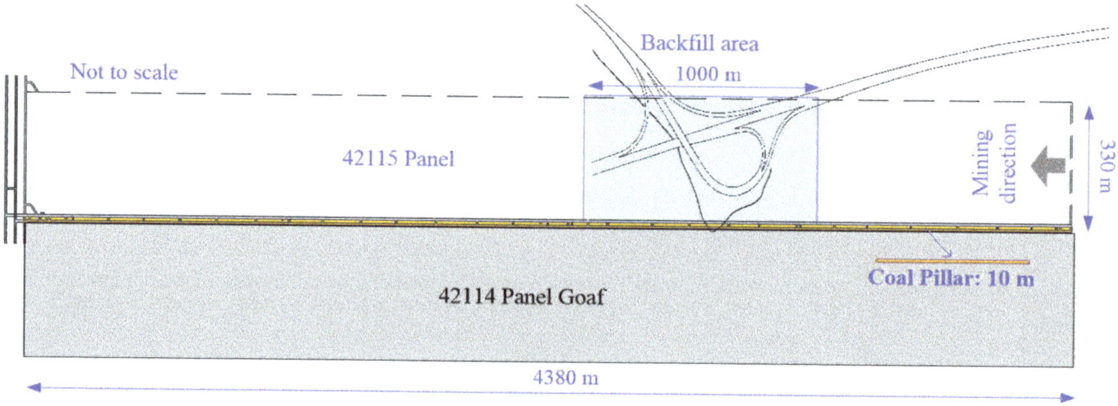

Figure 5. The relationship and relative location of the yield coal pillar, working face, interchange bridge and the structure backfill area.

Table 2. Initial roof support parameters in the backfill area.

Support Materials	Parameters	Requirements	Remarks
Rock bolts	Diameter/length Spacing and row spacing Tray Anchorage agent	Φ 18 mm × L 1800 mm 1000 mm, 800 mm 140 mm × 250 mm × 10 mm W-shaped steel tray MSCK 23/80 mm	Screw thread steel
Anchor cables	Diameter/length Spacing and row spacing Tray Anchorage agent	Φ 15.24 × L 7500 mm 2000 mm, 1600 mm 300 mm × 300 mm × 16 mm dished tray MSCK23/60 mm + MSZ23/80 mm	Mine anchor cables
Rhombic metal mesh	Specification Length Width	8# 2800 mm 1000 mm	Two pieces lapped by 100 mm
Hydraulic support	Specification Support height Center distance Rated working resistance Rated initial support force	ZY12000/20/40D 2000~4000 mm 1750 mm 12,000 kN 7917 kN	
Backfill hydraulic support	Specification Support height Center distance Rated working resistance Rated initial support force	ZZC10000/20/40 2000~4000 mm 1500 mm 10,000 kN 8322 kN	
Single hydraulic prop	Specification Maximum support height Rated working resistance Spacing	DW31.5-300/100 3500 mm 200 kN 1000 mm	

3. Theoretical Model of Backfill Parameters and Finite Element Simulation of Roof Subsidence

3.1. The Ultimate Collapse Step Distance of the Main Roof of Underground Structural Backfill

The study and application of structural backfill technology of goaf with high-water materials and pillars is an innovative approach to solve the problem of coal mining in 42,115 working face of Liangshuijing coal mine passing through the interchange bridge.

The high-water materials have the performance of fast loading, and it supports the main roof load together with the pillars. In the underground structural backfill area, when the filling barricades bear the immediate roof, the determination of the ultimate fracture step distance of the main roof is the key factor in studying the partition size of structural backfill.

3.1.1. Assessment of the Main Roof Load Using the Periodic Weighting Step Distance

According to the mine pressure observation of Liangshuijing coal mine, in the mining process of 42,114 working face, which is adjacent to 42,115 working face in the study area, the periodic weighting step distance is 15~19 m, with an average of 17 m. In the case of the traditional mining process of 42,115 working face, the main roof periodic weighting beam structure is a cantilever beam, as shown in Figure 6, the stress structure diagram and bending moment of the beam are as follows [38,39]:

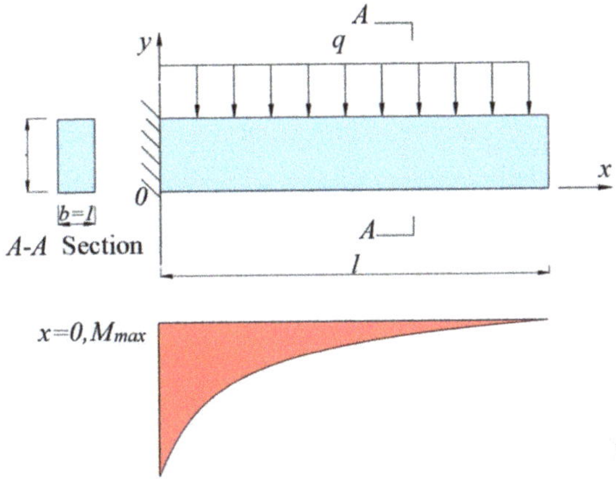

Figure 6. Force model and bending moment diagram of a cantilever beam.

According to the force and bending moment diagram, it can be determined that, at the point $x = 0$, the force F is:

$$F = -ql \tag{4}$$

where l is the length of the cantilever beam; q is the uniform load on the cantilever beam.

When $x = 0$, the maximum bending moment M is:

$$M = -\frac{1}{2ql^2} \tag{5}$$

Normal stress σ at any point in the cantilever beam is:

$$\sigma = -\frac{My}{I_z} \tag{6}$$

The maximum bending moment occurs at $x = 0$, $y = \pm h/2$, so:

$$\sigma_{\max} = -\frac{\frac{1}{2}ql^2\frac{h}{2}}{\frac{1}{12}h^3} = \frac{3ql^2}{h^2} \tag{7}$$

where $\sigma_{\max}$ is the maximum normal stress in the cantilever beam; h is the thickness of the cantilever beam.

When $\sigma_{\max} = R_T$, the rock layer's normal stress achieves its ultimate tensile strength, the rock layer splits at this point, and the ultimate span of beam fracture is:

$$L_{IT} = h\sqrt{\frac{R_T}{3q}} \tag{8}$$

Owing to the 42,115 working face's rock tensile strength $R_T = 1.9$ MPa, the periodic weighting step distance $L_{IT} = 17$ m, and the thickness of the main roof $h = 15$ m, so:

$$q = \frac{R_T h^2}{3{L_{IT}}^2} = \frac{1.9 \times 10^6 \times 15^2}{3 \times 17^2}\text{Pa} = 0.493\ \text{MPa} \tag{9}$$

3.1.2. Estimation of the Main Roof Fracture Step Distance Based on the Backfill Structure

Under backfill structure conditions, the periodic weighing step distance of the main roof is modelled as a mechanical structure with fixed-freely-supported, and the elastic-plastic bending of beams is assumed as follows [40,41]:

(1) The main roof material is ideal rigid plastic, that is to say, the rigid plastic deformation model is adopted without considering the elastic properties and strengthening effect of the material;

(2) Rock is a brittle material. Before failure, the deformation of the structure is small enough, and the geometric relationship of material deformation is linear;

(3) Before reaching the ultimate load, the structure will not lose its stability.

Figure 7 shows the stress analysis and bending moment diagram of fixed-freely-supported after backfilling with high-water materials in the structural backfill system.

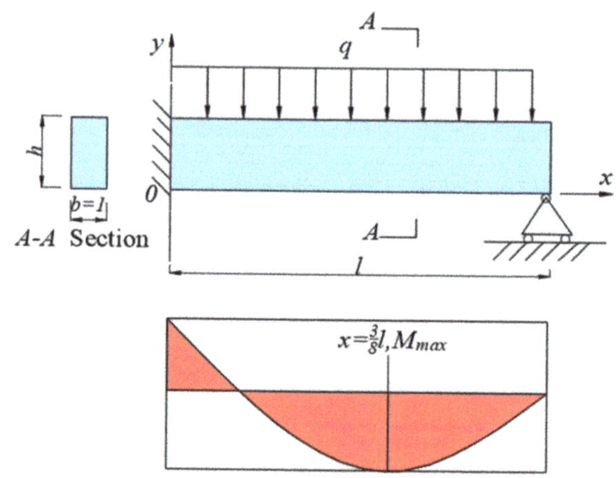

Figure 7. Force model and bending moment diagram with a fixed-freely-supported beam.

According to the principles of structural mechanics and elastic mechanics:

$$F_0 \Delta 1l + \Delta 1p = 0 \tag{10}$$

In addition:

$$\Delta 1l = \frac{1}{3}l^3\left(\frac{l}{EI}\right), \Delta 1p - \frac{1}{8}ql^4\left(\frac{l}{EI}\right) \tag{11}$$

Then:

$$F_0 = -\frac{3}{8}ql, F_l = -\frac{5}{8}ql \tag{12}$$

It can be determined according to the bending moment diagram: when $x = 3l/8$, the bending moment M is the maximum, $-9ql^2/128$.

Normal stress σ at any point in the fixed-freely-supported beam is:

$$\sigma = -\frac{My}{I_z} \tag{13}$$

The maximum bending moment occurs at $x = 3l/8, y = \pm h/2$, so:

$$\sigma_{\max} = -\frac{\frac{9}{128}ql^2\frac{h}{2}}{\frac{1}{12}h^3} = \frac{27ql^2}{64h^2} \tag{14}$$

When $\sigma_{\max} = R_T$, the rock layer's normal stress develops its ultimate tensile strength, the rock layer splits at this point, and the ultimate span of beam fracture under the condition of the support of the wall filled with the underground structural filling high-water materials are:

$$L_{IT} = 8h\sqrt{\frac{R_T}{27q}} \tag{15}$$

According to the model result above, the main roof load q is equal to 0.493 MPa, so the ultimate span of beam fracture $L_{IT} = 45.33$ m, and considering the safety factor of 2.0, the backfill step distance is 22.7 m.

3.1.3. The Procedure of Resolving Beam Load Problems

The beam load is determined using composite beam theory. The immediate roof, the main roof, and the strata of each stratum comprise the upper portion of the coal seam. It must meet the uniform load of the overlaying strata in this location [38,42]. As illustrated in Figure 8, there are m layers of strata above the backfill barricade, with the elastic modulus, thickness, and volume force of each layer being $E_i(i = 1,2, \ldots, m)$, $h_i(i = 1,2, \ldots, m)$, and $\gamma_i(i = 1,2, \ldots, m)$, respectively.

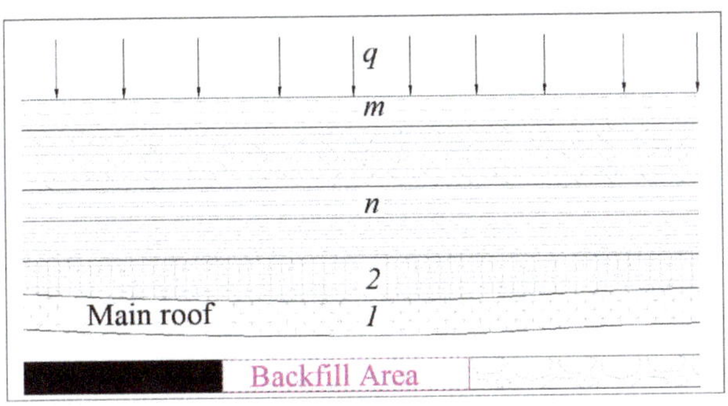

Figure 8. Solving model of composite beam roof load.

Assume that the main roof is the first layer, with n strata controlled by it. The n layers maintain contact, and the lower layer's deformation governs the top layer's deformation. Together, the n layers create a composite beam. According to composite beam theory, the rock section's bending moment M and shear force Q fulfil the following equation:

$$\begin{cases} Q = Q_1 + Q_2 + \cdots + Q_n \\ M = M_1 + M_2 + \cdots + M_n \end{cases} \tag{16}$$

In the case of structural backfill mining, the subsidence deformation of the composite beam is small, and the curvature of each rock stratum in the composite beam is consistent after deformation:

$$k_i = \frac{1}{\rho_i} = \frac{(M_i)_x}{E_i I_i} \tag{17}$$

where ρ_i is the curvature radius of layer i, E_i is the elastic modulus of layer i, and I_i is the moment of inertia of the i stratum.

$$\frac{M_1}{E_1 I_1} = \frac{M_2}{E_2 I_2} = \cdots = \frac{M_n}{E_n I_n} \tag{18}$$

That is:

$$\frac{(M_1)_x}{(M_2)_x} = \frac{E_1 I_1}{E_2 I_2}, \frac{(M_1)_x}{(M_3)_x} = \frac{E_1 I_1}{E_3 I_3}, \cdots, \frac{(M_1)_x}{(M_n)_x} = \frac{E_1 I_1}{E_n I_n} \tag{19}$$

In addition:

$$M_x = (M_1)_x + (M_2)_x + \cdots + (M_n)_x = (M_1)_x \left(1 + \frac{E_2 I_2 + E_3 I_3 + \cdots + E_n I_n}{E_1 I_1}\right) \tag{20}$$

The shear force is the first derivative of the bending moment, so:

$$\frac{dM}{dx} = Q \tag{21}$$

That is:

$$(Q_1)_x = \frac{E_1 I_1}{E_1 I_1 + E_2 I_2 + \cdots E_n I_n} \cdot Q_x \tag{22}$$

The beam load is the first derivative of the shear force, so,

$$\frac{dQ}{dx} = q \tag{23}$$

That is:

$$(q_1)_x = \frac{E_1 I_1}{E_1 I_1 + E_2 I_2 + \cdots E_n I_n} \cdot q_x \tag{24}$$

In addition:

$$\begin{cases} q_x = \gamma_1 h_1 + \gamma_2 h_2 + \cdots + \gamma_n h_n \\ I_1 = \frac{b h_1^3}{12}, I_2 = \frac{b h_2^3}{12}, \cdots, I_n = \frac{b h_n^3}{12} \end{cases} \tag{25}$$

$(q_1)_x$ is the load formed when considering the influence of n layers on the first layer, which is $(q_n)_1$, then:

$$(q_n)_1 = \frac{E_1 h_1^3 (\gamma_1 h_1 + \gamma_2 h_2 + \cdots + \gamma_n h_n)}{E_1 I_1 + E_2 I_2 + \cdots E_n I_n} \tag{26}$$

The roof load of the filling barricade in the backfill structure, Q, is:

$$Q = q_1 + q_2 = 0.618 \text{ MPa} \tag{27}$$

According to the calculation of coal seam buried depth, the permanent load that the filling barricade needs to bear is 3.75 MPa (permanent load).

3.2. Subsidence Analysis of Main Roof and Immediate Roof Using Finite Element Model

According to the rock drilling histogram in Section 2, the thickness of the overlying immediate roof is 5 m, and the thickness of the floor is 10 m. The full-scale model was established by Abaqus [39] as shown in Figure 9, and the dimensions of the immediate roof and floor are adjusted with the spacing of the supporting barricades, and the corresponding parameters of the rock stratum are defined; Hexahedral element (HEX) is adopted for rock stratum and pillars, consolidation constraint is adopted for wall bottom, hard contact force is adopted for barricade and immediate roof and floor, and tie constraint is applied for relative space position. The subject of this section is the influence of different filling parameters on roof subsidence. The main factors considered are the influence of temporary load and permanent load on roof subsidence. The influence of immediate roof and main roof mechanical parameters has exceeded the scope of this paper. Therefore, in the simulation process, the immediate roof and the main roof are simplified to the same material, called the "upper rock layer", with a thickness of 20 m. The mechanical parameters applied in the model are shown in Table 2. The dimension of the "upper rock layer" is 330 m in length and 20 m in height, with mesh of 1 m in all three directions. The applied surface load has been calculated to be 0.618 MPa (temporary load) and 3.75 MPa (permanent load).

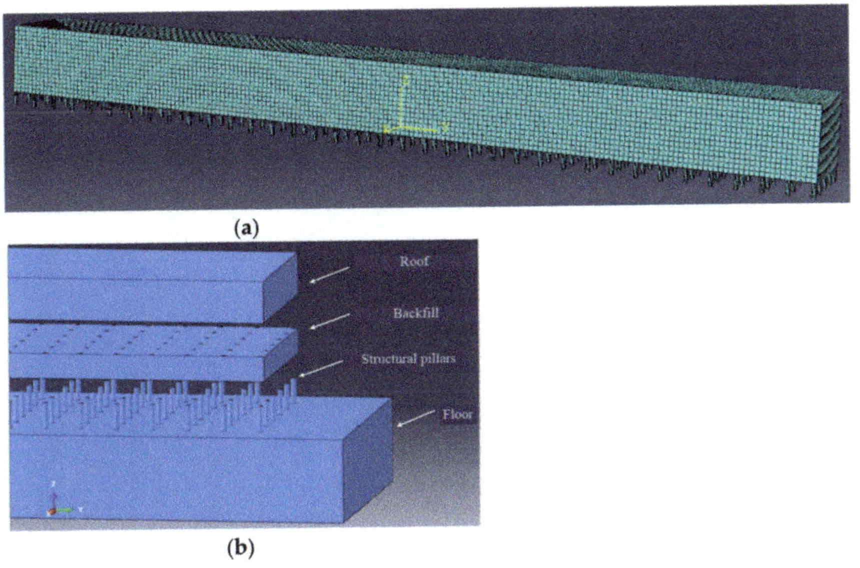

Figure 9. Finite element model of structural pillars backfill. (**a**) Global model diagram; (**b**) Local component diagram.

In the X positive and negative direction of the coordinate system, the filled partition barricade model sets the position for the partition barricade, and the mechanical behavior of directly jacking in these areas completely depends on the support effect of the barricade, so there is no constraint condition. So, its deformation behavior can only be obtained through the constraint effect of the barricade. In the Y positive and negative direction of the coordinate system, the negative direction of the Y coordinate is the coal wall, which is set as a fixed constraint. The square of the Y coordinate is the direction of the connecting roadway, which can be different according to different reinforcement schemes, so there is no constraint.

According to the method in GB50936-2014 [43], GB50017-2017 [44], the ultimate bearing capacity of the precast concrete-filled steel tube pillars is calculated, and the relevant parameters of the Φ 500 mm L 4000 mm pillars under the temporary load of 0.618 MPa are shown in Table 3.

Table 3. Parameters of Φ 500 mm L 4000 mm pillars in structural filling.

Materials	Strength Grade	Diameter	Length	Ultimate Bearing Capacity	Spacing	Row Spacing
Concrete	C35	500 mm	4000 mm	10,200 kN	4	4

Through the above model, the finite calculation of the temporary wall support is carried out, and the deformation law of the roof is analyzed when the support wall spacing is 5 m, 10 m, 15 m, and 20 m respectively.

Through the comparative analysis of the subgraphs in Figure 10, the maximum settlement of the roof is in the center of the two wall supports in the X direction, and the vertical deformation of the wall support area and the wall is small, which also conforms to the deformation law of the unidirectional slab under uniform load. When only temporary load (0.618 MPa) is considered, the maximum settlement of direct roof with spacing of 5 m, 10 m, 15 m, and 20 m is 2.584 mm, 7.787 mm, 23.32 mm, and 63.24 mm, respectively. Considering the permanent support load (3.75 MPa), the maximum settlement of the roof with the spacing of 5 m, 10 m, 15 m, and 20 m is 15.68 mm, 47.25 mm, 144.6 mm, and 598.03 mm, respectively.

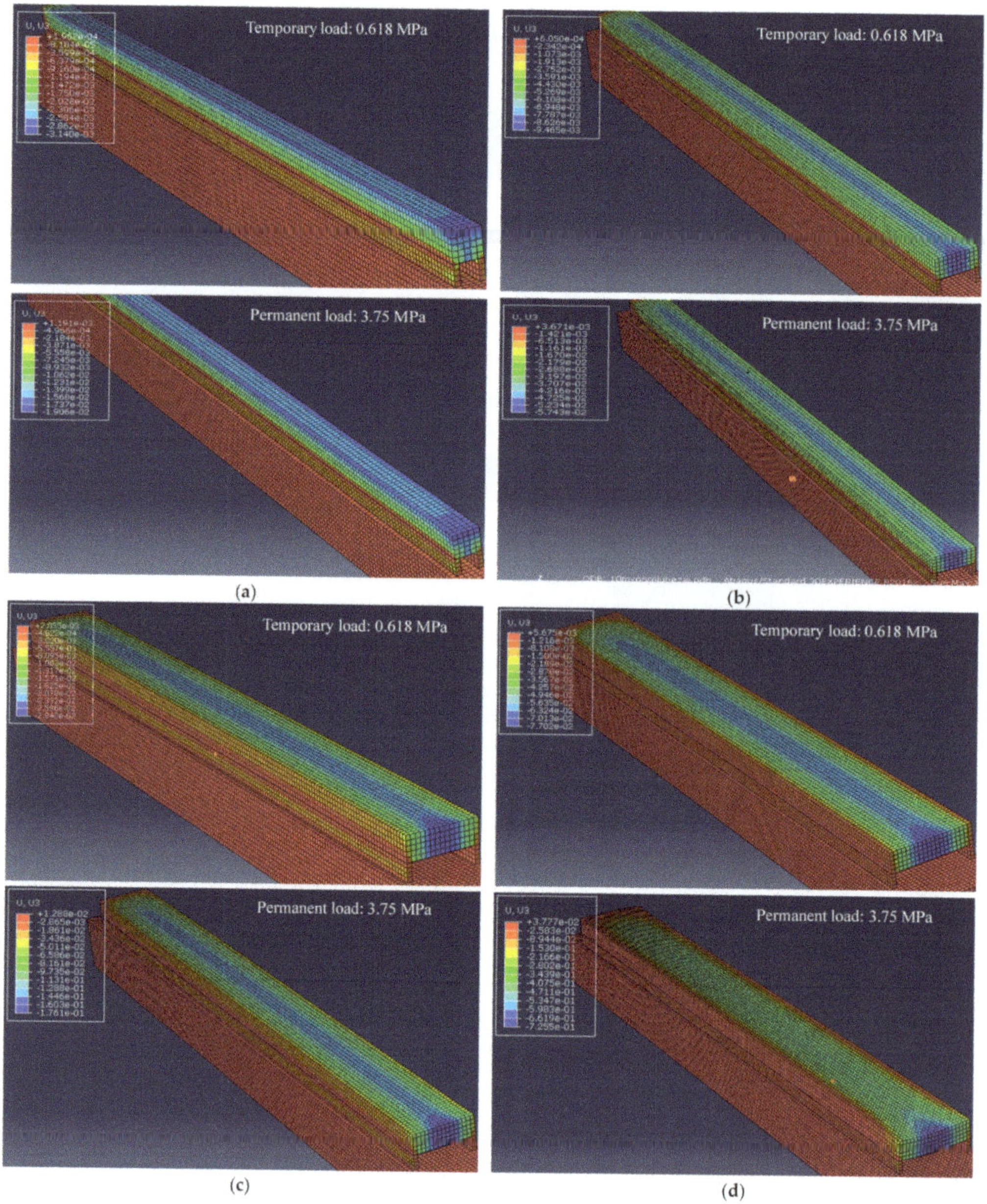

Figure 10. Vertical displacement cloud distribution map under different support barricade spacing. (**a**) Support barricade spacing is 5 m; (**b**) Support barricade spacing is 5 m; (**c**) Support barricade spacing is 15 m; and (**d**) Support barricade spacing is 20 m.

The numerical variation law of the maximum settlement area of the roof is shown in Figure 11. The settlement rate of the immediate roof increases with the increase of the spacing between the supporting barricades. The settlement rate of the immediate roof increases more rapidly in the section of 15~20 m, and the settlement rate of the immediate roof increases more gently in the section of 5~10 m. The maximum displacement calculated by the column model with boundary constraints is 8.375 mm, while the maximum settlement calculated by the 5 m and 10 m infilled pillar support scheme is 2.584 mm and 7.787 mm, which shows that the infilled wall construction scheme is more reasonable. The infilled wall restricts the roof settlement to a certain extent, the construction is more convenient, and the cost is relatively low.

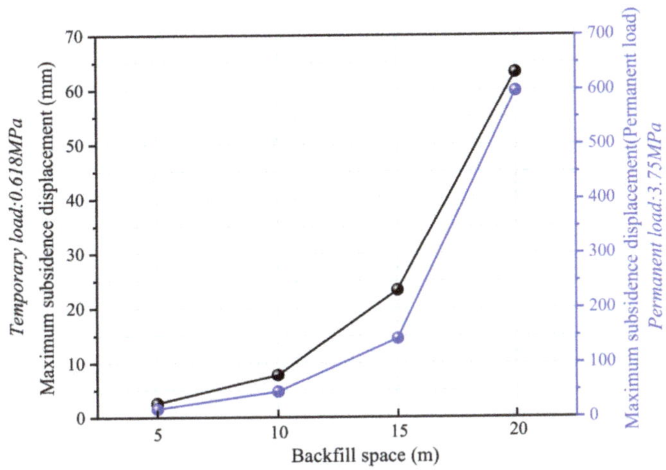

Figure 11. Wall spacing-maximum settlement of the roof.

3.3. Determine the Structure Backfill Mining Parameters

(1) Material and size selection of pillars

The pillars support the roof strata of the backfill area. Related parameters such as pillar size and materials are discussed in Section 3.2. There are mainly three commonly used concrete types for the materials of the pillars, namely C35, C40, and C60. In the structural backfill method, the column and the partition wall are supported together, and the bearing capacity of the column is not the most crucial factor. Therefore, the commonly used C35 concrete is selected. There are mainly two commonly used diameters for pillars, namely Φ 800 mm and Φ 500 mm. According to the calculation method of bearing capacity provided by GB50936-2014 [43], GB50017-2017 [44], the parameters of different types of pillars under the temporary load of 0.618 MPa are shown in Table 4. Φ 500 mm, although the bearing capacity provided is less than Φ 800 mm, because its weight is one-third of Φ 800 mm, the cost is one-half of Φ 800 mm. Therefore, a Φ 500 mm pillar prefabricated with C35 concrete was chosen.

(2) Material and size selection of barricade

Regarding the material of the filling barricade, concrete, sand, and gravel can be selected [45,46]. The selection of high-water materials has advantages over other materials due to its fast load-bearing characteristics. The mechanical load-bearing properties of high-water materials are discussed in detail in Section 5.1 and are also proven in previous literature [47,48]. In Section 3.2, the variation relationship of the roof subsidence of different backfill barricades spacing is simulated, and the roof subsidence is smaller under the filling barricade spacing of 5~15 m, considering the spacing and row spacing between the columns are both 4 m. Therefore, the spacing of the filling barricade is determined to be 12 m. In addition, the width of the filling barricade is 1.5 m, the maximum distance between the filling barricade and the pillars and the coal mining face is 22.7 m (discussed in Section 3.1.2).

Table 4. The parameters of different types of pillars.

Materials	Strength Grade	Diameter (mm)	Length (mm)	Ultimate Bearing Capacity (kN)	Spacing (m)	Row Spacing (m)	Cost * (CNY)	Weight (kg/each)
Concrete	C60	800	4000	28,150	6.5	7	3900	4040
Concrete	C40	800	4000	24,500	6	6.5	3860	4006
Concrete	C35	800	4000	23,000	4	4	3830	4001
Concrete	C60	500	4000	11,500	4	4.5	1880	1660
Concrete	C40	500	4000	10,900	4	4.4	1870	1643
Concrete	C35	500	4000	10,200	4	4	1850	1641

Cost *—The value is the project cost of the pillars, which is different from the cost in the Section 5.3 (considering the cost of surface transportation and underground transportation) of the calculation of economic benefits.

4. Prediction Model of Mining Deformation and Subsidence and Its Application

4.1. Basic Theory of Prediction Model

The probability integral theory hypothesizes that the law of strata and surface movement induced by mining is comparable to the law given in macroscopic terms by the granular medium model as a randomized medium, and it has developed into one of the most matured and extensively used prediction methods [49–51].

As shown in Figure 12, unit $B(u,v)$ is mined in an inclined coal seam by taking ground coordinate system xOy and coal seam coordinate system uO_yv. According to the principle of probability integral method, the subsidence value of surface point $A(x,y)$ caused by unit mining is:

$$W_e(x,y) = \frac{1}{r^2}\exp\left(-\pi\frac{(x-u)^2+(y-v+H\cdot c\tan\theta)^2}{r^2}\right) \tag{28}$$

where r is the main influence radius at unit B, $r = H/\tan\beta$; H is the mining depth at unit B; tanβ is the tangent of the main influence angle; θ is the propagation angle of mining influence.

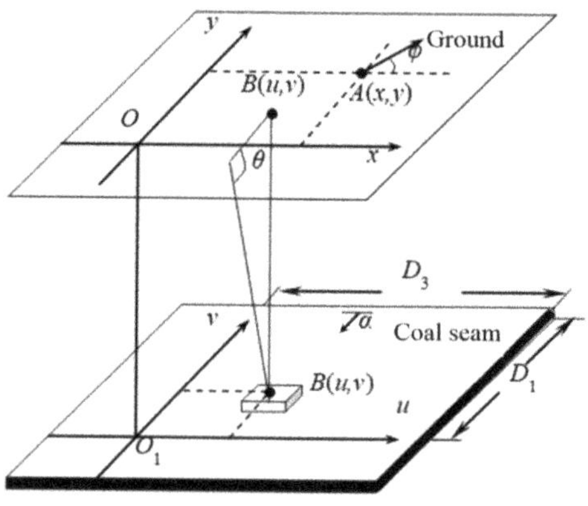

Figure 12. Space mining coordinate system.

Accordingly, the inclination $(i_{ex}(x,y), i_{ey}(x,y))$, curvature $(K_{ex}(x,y), K_{ey}(x,y))$, horizontal movement $(U_{ex}(x,y), U_{ey}(x,y))$ and horizontal deformation $(\varepsilon_{ex}(x,y), \varepsilon_{ey}(x,y))$ of surface point $A(x,y)$ along x, y directions are as follows:

$$\left.\begin{array}{l} i_{ex}(x,y) = -\frac{2\pi(x-u)}{r^2} W_e(x,y) \\ i_{ey}(x,y) = -\frac{2\pi(y-v+H\cdot c\tan\theta)}{r^2} W_e(x,y) \end{array}\right\} \quad (29)$$

$$\left.\begin{array}{l} K_{ex}(x,y) = -\frac{2\pi}{r^2}\left(1 - \frac{2\pi(x-u)^2}{r^2}\right) \cdot W_e(x,y) \\ K_{ey}(x,y) = -\frac{2\pi}{r^2}\left(1 - \frac{2\pi(y-v+H\cdot c\tan\theta)^2}{r^2}\right) W_e(x,y) \end{array}\right\} \quad (30)$$

$$\left.\begin{array}{l} U_{ex}(x,y) = b \cdot r \cdot i_{ex}(x,y) \\ U_{ey}(x,y) = b \cdot r \cdot i_{ey}(x,y) + W_e(x,y) \cdot c\tan\theta \end{array}\right\} \quad (31)$$

$$\left.\begin{array}{l} \varepsilon_{ex}(x,y) = b \cdot r \cdot K_{ex}(x,y) \\ \varepsilon_{ey}(x,y) = b \cdot r \cdot K_{ey}(x,y) + i_{ey}(x,y) \cdot c\tan\theta \end{array}\right\} \quad (32)$$

where b is the horizontal displacement coefficient. According to the superposition principle, the subsidence $W(x,y)$ of surface point $A(x,y)$ caused by the whole coal seam mining is:

$$W(x,y) = W_0 \int_{s_3}^{l} \int_{s_1 \cos\alpha}^{L} W_e(x,y)\,dv\,du \quad (33)$$

where W_0 is the maximum subsidence, $W_0 = mq\cos\alpha$; m is the normal mining thickness of coal seam; q is the subsidence coefficient; α is the dip angle of coal seam;

$$\left\{\begin{array}{l} l = D_3 - s_4 \\ L = (D_1 - s_2)\cos\alpha \end{array}\right. \quad (34)$$

where D_3 is the strike length of working face; D_1 is the inclined length of working face; s_1, s_2, s_3, s_4 is the inflection point offset of the left and right boundary in downhill direction, uphill direction and strike direction, respectively.

On this basis, it can be deduced that the surface point $A(x,y)$ along any direction φ, the inclination $i(x,y,\varphi)$, curvature $K(x,y,\varphi)$, horizontal movement $U(x,y,\varphi)$ and horizontal deformation $\varepsilon(x,y,\varphi)$ of the model are as follows:

$$i(x,y,\varphi) = W_0 \int_{s_3}^{l} \int_{s_1 \cos\alpha}^{L} (i_{ex}(x,y)\cos\varphi + i_{ey}(x,y)\sin\varphi)\,dv\,du \quad (35)$$

$$K(x,y,\varphi) = W_0 \int_{s_3}^{l} \int_{s_1 \cos\alpha}^{L} \left(K_{ex}(x,y)\cos^2\varphi + K_{ey}(x,y)\sin^2\varphi + \frac{4\pi^2(x-u)(y-v+H\cdot c\tan\theta)}{r^4} W_e(x,y)\sin 2\varphi\right) dv\,du \quad (36)$$

$$U(x,y,\varphi) = W_0 \int_{s_3}^{l} \int_{s_1 \cos\alpha}^{L} (U_{ex}(x,y)\cos\varphi + U_{ey}(x,y)\sin\varphi)\,dv\,du \quad (37)$$

$$\varepsilon(x,y,\varphi) = W_0 \int_{s_3}^{l} \int_{s_1 \cos\alpha}^{L} \left(\varepsilon_{ex}(x,y)\cos^2\varphi + \varepsilon_{ey}(x,y)\sin^2\varphi + \left(\frac{8b\pi^2(x-u)(y-v+H\cdot c\tan\theta)}{r^3} \cdot W_e(x,y) + i_{ex}(x,y)c\tan\theta\right)\sin\varphi\cos\varphi\right) dv\,du \quad (38)$$

When the surface is fully mined, the maximum deformation can be calculated by the following equation: Maximum value of ground surface subsidence, $W_0 = mq\cos\alpha$, inclination, $i_0 = W_0/r$, curvature, $k_0 = \pm 1.52W_0/r^2$, horizontal movement, $U_0 = bW_0$, and horizontal deformation, $\varepsilon_0 = \pm 1.52bW_0/r$.

4.2. Input Parameters of the MSAS

The Mining subsidence prediction and analysis system (MSAS) was developed on the basis of the visual basic 6.0 system, which can truly realize the prediction calculation of surface subsidence, inclination, curvature, horizontal movement and horizontal deformation caused by mining of any shape working face [52]. In addition, the system has the advantages of simple data preparation and convenient operation, which abandons the traditional method of rectangular partitions for any working face, and adopts the integral method to calculate, which reduces the partition error and makes the calculation results more accurate. Consequently, the contour lines of maximum inclination, maximum curvature and maximum horizontal deformation can be given, which makes the analysis of mining damage easier. Considering the reasonable length of the article, there is no detailed study on the expected parameters, and the relevant parameters in the reference [53] are selected when the prediction model is applied, as follows:

(1) Normal mining prediction parameters of the MSAS:

Surface subsidence coefficient, $q = 0.72$, horizontal movement coefficient, $b = 0.3$, main influence angle tangent, $\tan\beta = 1.8$, inflection point translation distance, $S_0 = 0.05H$, mining influence spread angle, $\theta = 90 - 0.5\alpha$.

(2) Backfill mining prediction parameters of the MSAS:

Surface subsidence coefficient, $q = 0.085$, horizontal movement coefficient, $b = 0.3$, main influence angle tangent, $\tan\beta = 1.8$, inflection point translation distance, $S_0 = 0.03H$, mining influence spread angle, $\theta = 90 - 0.5\alpha$.

4.3. Model Application and Result Analysis

Through the calculation of the MSAS system, the surface subsidence and deformation of the interchange bridge area after 42,115 working face mining can be obtained. The contour map of surface subsidence and the contour map of maximum inclination, curvature and horizontal deformation are shown in Figures 13 and 14.

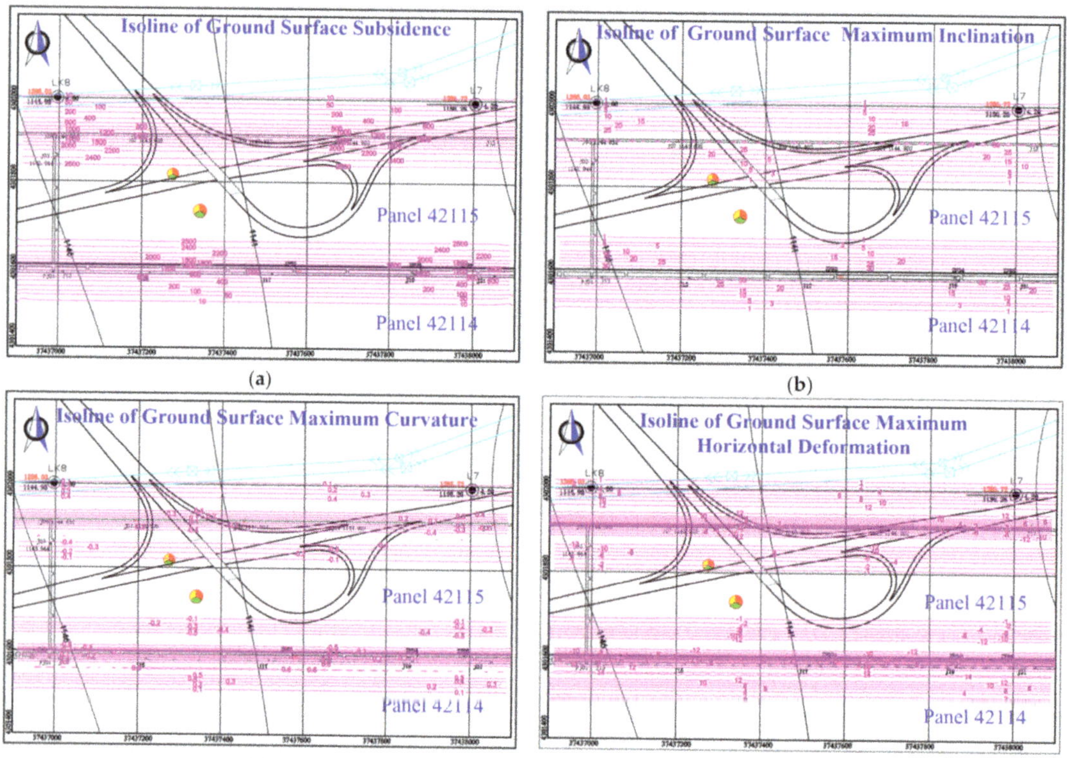

Figure 13. Prediction results of surface deformation after 42,115 working face normal mining. (**a**) Ground surface subsidence; (**b**) Ground surface inclination; (**c**) Ground surface curvature; (**d**) Ground surface horizontal deformation.

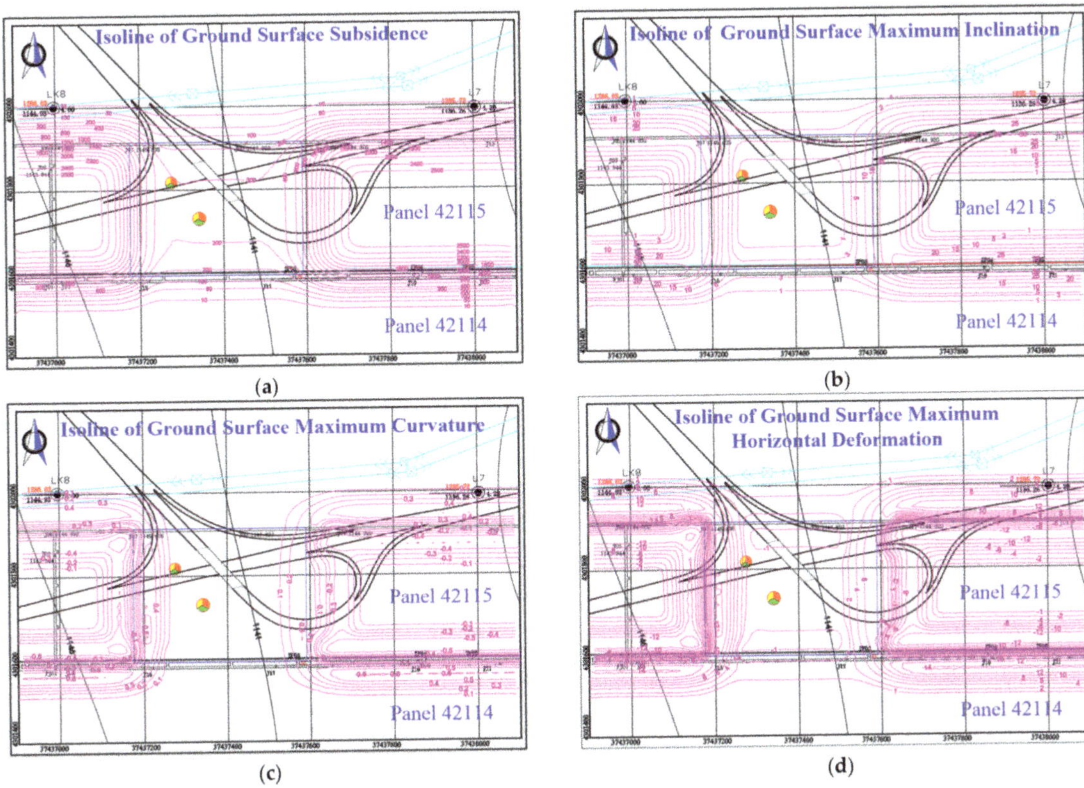

Figure 14. Prediction results of surface deformation after 42,115 working face backfill mining. (**a**) Ground surface subsidence; (**b**) Ground surface inclination; (**c**) Ground surface curvature; (**d**) Ground surface horizontal deformation.

The interchange bridge area final subsidence and maximum deformation values of normal mining and backfill mining are shown in Figures 13–15. Under normal mining conditions, the maximum subsidence at the north end of the interchange bridge area is 2300 mm, and the maximum subsidence at the south end of the interchange bridge area is 2550 mm. Under backfill mining conditions, the maximum subsidence at the north end of the interchange bridge area is 240 mm, and the maximum subsidence at the south end of the interchange bridge area is 300 mm. Under normal mining conditions, the maximum inclination of the north end of the interchange bridge area is 15 mm/m, and under backfill mining conditions, the maximum inclination of the north end of the interchange bridge area is 1.8 mm/m. Under normal mining conditions, the maximum curvature of the north end of the interchange bridge area is -0.4 mm/m^2, and under backfill mining conditions, the maximum curvature of the north end of the interchange bridge area is -0.1 mm/m^2. Under normal mining conditions, the maximum horizontal deformation of the north end of the interchange bridge area is -12 mm/m, and under filling mining conditions, the maximum curvature of the north end of the interchange bridge area is -1.5 mm/m.

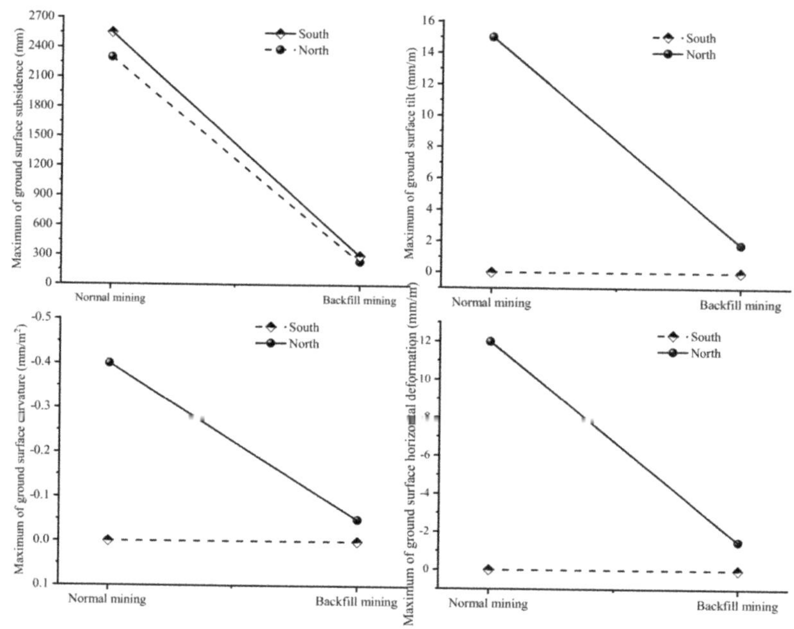

Figure 15. Final value of surface subsidence prediction for normal mining and backfill mining.

Due to the interchange bridge being located directly above the 42,115 working face, which is located in the center of the subsidence basin after mining, each deformation value is small. During the mining process, the interchange bridge is affected by various dynamic deformations, especially when the working face is pushed under the interchange bridge, then each deformation value is the largest. The interchange bridge is first subjected to positive curvature and tensile deformation, and then to negative curvature and compression deformation. For this reason, the surface subsidence deformation of 42,115 working face under the interchange bridge area is specially predicted. The maximum dynamic deformation value of normal mining and backfill mining under the interchange bridge area is shown in Table 5.

Table 5. Dynamic maximum evolution value of surface subsidence prediction between normal mining and backfill mining.

Mining Method	Location	Tilt (mm/m)	Curvature (mm/m²)	Horizontal Deformation (mm/m)
Normal mining	South	30	−0.5~0.5	−14~14
	North	25	−0.4~0.4	−12~12
Backfill mining	South	3.5	−0.05~0.05	−1.5~1.5
	North	3.0	−0.05~0.05	−1.3~1.3

As shown in Table 5, at the south of the interchange bridge, the maximum dynamic tilt, curvature, and horizontal deformation under normal mining conditions are 30 mm/m, -0.5~0.5 mm/m^2, and -14~14 mm/m compared to backfill mining, which are 3.5 mm/m, -0.05~0.05 mm/m^2, and -1.5~1.5 mm/m. At the north of the interchange bridge, the maximum dynamic tilt, curvature, and horizontal deformation under normal mining conditions are 25 mm/m, -0.4~0.4 mm/m^2, and -12~12 mm/m compared to backfill mining which are 3.0 mm/m, -0.05~0.05 mm/m^2, and -1.3~1.3 mm/m. Obviously, the surface subsidence parameters are reduced appropriately by ten times under the conditions of backfill mining. In other words, backfill mining can effectively limit surface deformation and greatly reduce the impact of mining on surface buildings, which is consistent with the previous research result [4,16,54–56].

5. Discussions

5.1. Feasibility of Structural Backfill Engineering

Several parameters govern the mechanical characteristics of backfill materials, which could be divided into two categories: macroscopic parameters and intrinsic parameters [3,57]. Ingredients A and B make up the high-water materials. Ingredient A is a unique concrete that contains the minerals listed in Table 6. The major components of ingredient B are quicklime (CaO) and gypsum ($CaSO_4$). Ingredients A and B are combined with water and fluttered constantly for 24 h to keep the mixture uncondensed. Ingredients A and B are combined with water and fluttered constantly for 24 h to keep the mixture uncondensed. The major components of high water material are ettringite, alumina, and free water [47], which has the benefit of quick healing and a significant post peak deformation with a high load carrying capacity. The full stress-strain curve and deformation characteristics of high-water materials after 7 days of setting at the water cement ratio of 1.5:1 is shown in Figure 16 [58–60].

Table 6. Mineral composition of ingredient A.

Mineral Composition	$3CaO{\cdot}3Al_2O_3{\cdot}CaSO_4$	$2CaO{\cdot}SiO_2$	$4CaO{\cdot}Al_2O_3{\cdot}Fe_2O_3$
Content/%	3~13	30~48	1~3

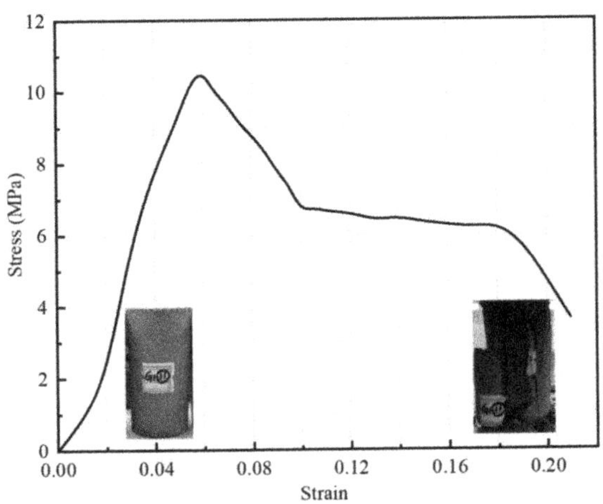

Figure 16. Full stress strain curve of high-water materials.

It can be seen from Figure 16 that high-water materials have outstanding plastic characteristics. After the load reaches the peak strength, the high-water materials do not completely destroy and lose the bearing capacity immediately. However, with the further increase of strain, the bearing capacity decreases slowly, and the rate of bearing capacity decreasing slowly with the increase of strain is far less than that of ordinary concrete and rock materials. Under this condition, the peak compressive strength of high-water materials is 10.4 MPa. When the strain reaches 0.1, the compressive strength is still above 65% of the peak strength, the strain continues to increase to 0.18, and the residual compressive strength is about 59% of the peak strength [61]. The barricade of high-water materials shows obvious plastic materials characteristics, which can allow large plastic deformation under pressure, and the strength attenuation is relatively slow, which can maintain high residual strength.

5.2. Measured Analysis of Support Resistance

After adopting the structural backfill and mining technology, the monitoring result of the hydraulic support resistance is shown in Figure 17 when the 42,115 working face passes through the interchange bridge region.

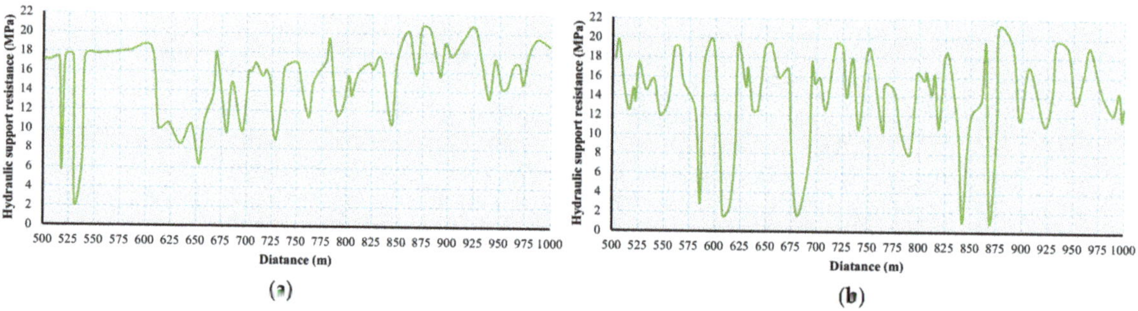

Figure 17. Curve of the support resistance of 42,115 working face under the interchange bridge area. (**a**) Resistance of 80 # hydraulic support; (**b**) Resistance of 90# hydraulic support.

As shown in Figure 17, the overall resistance of hydraulic support is less than 22 MPa, and there is no obvious strata behavior due to the supporting effect of filling body on the roof of 42,115 working face. The resistance of the hydraulic support in the backfill area of the 42,115 working face is relatively small, indicating that the small roof pressure of the working face verifies the rationality of the structural backfill parameters, and it is also consistent with the surface subsidence results predicted in Section 4.3.

5.3. Economic and Social Benefit Analysis

(1) Economic benefit analysis:

The output value of coal mine mining by using the structure backfill method:

The length and width of the mining area affected by the interchange bridge are 1000 m and 330 m respectively. After adopting the structural backfill technology of high-water materials combined with pillars for this region, the increased output value of coal mine mining can be calculated by the following equation:

$$G = LDm\eta\gamma V \tag{39}$$

where G is the increase output value of 42,115 working face, RMB; L is the increase length of 42,115 working face, 1000 m; D is the increase width of 42,115 working face, 330 m; m is the coal seam thickness of 42,115 working face, 3.5 m; η is the recovery rate of 42,115 working face, 95%; γ is the coal seam bulk density of 42,115 working face, 1.29 t/m^3; and V is the price per ton of coal, 600 CNY, so:

$$G = 1000 \times 330 \times 3.5 \times 95\% \times 1.29 \times 600 = 849{,}271{,}500 \text{ CNY} \tag{40}$$

The cost of using the structure backfill method is shown in Table 7:

Table 7. The cost of using the structure backfill method.

	Item	Sum Number	Price	Cost/CNY
Materials	Rock bolts	2300	36 CNY/series *	82,800
	Anchor cable	500	135 CNY/series *	67,500
	Metal mesh	2000 m^2	17 CNY/m^2	34,000
	High-water materials	2700 t	2150 CNY/t	5,805,000
Equipment	Backfill hydraulic support	40	650,000 CNY/each	-
	Hydraulic support	40	486,000 CNY/each	-
	Difference between hydraulic support	-	16,400 CNY/each	65,600
	Two-fluid grouting pump	2	350,000 CNY/each	700,000
	Mixing drum	4	25,000 CNY/each	100,000
Pillars	Precast concrete column	11,000	2100 CNY/each	23,100,000
Backfill working face	Worker	1320 days	300 CNY/day	396,000
	High-water materials	66,000 t	2150 CNY/t	141,900,000
	Sum			172,250,900

* A series of rock bolt contains the anchorage agent and tray; A series of anchor cable contains the anchorage agent, anchor lock and tray.

Therefore, the economic benefit after using the structure backfill mining method is:
Economic benefit = 849,271,500 − 172,250,900 = 677,020,600 CNY

(2) Social benefit analysis:

This technology has the potential to significantly improve resource recovery rates, noticeably reduce coal resource waste, extend the life of the mine, and ensure the establishment of a high-yielding and efficient mine [46,62]. This technology not only solves the problem of safe mining of coal resources under the overpass, but also deals with a large amount of underground discharge and accumulation of gangue, alleviating the social problems such as no gangue discharge space and environmental pollution. The successful implementation of mixed filling mining of gangue and structural column widens the application scope of filling mining method, which provides an important engineering reference for coal mining under similar conditions.

In the mining process of 42,115 working face, the structural backfill technology of high-water materials combined with pillars for goaf is adopted. After mining, the height of roof crack development and vertical displacement are small, which can ensure the stability of the interchange bridge above the 42,115 working face and meet the requirement of safe production [63]. It not only achieves significant economic benefits, but also has considerable social benefits.

6. Conclusions

In this paper, through field measurement, theoretical analysis, and numerical simulation, a structural backfill method based on roof structural characteristics is proposed. The main conclusions obtained are as follows:

(1) Based on the monitoring result of the hydraulic support resistance, the periodic weighting step is determined to be an average of 17 m, then the main roof load is calculated to be to be 0.493 MPa, and the backfill step distance is determined to be 22.7 m. According to the composite beam theory of roof strata, the temporary load of the backfill structure is 0.618 MPa, and the permanent load is 3.75 MPa.

(2) The finite element model is established, then the roof layer mechanical properties and pillars parameters is selected depending on the related criteria. The result of the finite element model shows that the subsidence is lower when the spacing of the filling barricade is 5~15 m, considering the pillars layout, the reasonable filling spacing (12 m) of the barricade is determined, and the rationality of backfill mining for safe mining under the interchange bridge is verified by the probability integral method result.

(3) This paper discusses the feasibility of the structural backfill technology with high-water materials and pillars in goaf, and finally analyses the economic benefits of

677,020,600 CNY and demonstrates the application value of the structural backfill mining method based on the roof structural characteristics and its subsidence prediction model.

The structural filling strategy proposed in this paper with high-water materials and pillars is of great significance to solving the mining face under the building. It has significant advantages with the rapid bearing of high-water materials and the stable support of the pillars, which provides an innovation in the field of backfilling mining.

Author Contributions: Conceptualization, S.L. and J.B.; methodology, X.W.; software, S.L., G.W.; validation, S.L., J.B., X.W., G.W. and B.W.; formal analysis, G.W.; writing—original draft preparation, S.L., J.B. and G.W.; writing—review and editing, S.L., J.B., X.W., G.W. and B.W.; funding acquisition, J.B. and S.L. All authors have read and agreed to the published version of the manuscript.

Funding: This work was supported by the Postgraduate Research & Practice Innovation Program of Jiangsu Province, grant number KYCX21_2388. This research was also funded by the National Natural Science Foundation of China, grant number No. 52074239.

Data Availability Statement: Data is available on request from the corresponding author of the manuscript.

Acknowledgments: The authors are grateful to the reviewers for their helpful advice and comments.

Conflicts of Interest: The authors declare no conflict of interest.

References

1. Yang, Z.F.; Li, Z.W.; Zhu, J.J.; Yi, H.W.; Hu, J.; Feng, G.C. Deriving Dynamic Subsidence of Coal Mining Areas Using InSAR and Logistic Model. *Remote Sens.* **2017**, *9*, 125. [CrossRef]
2. Zheng, M.N.; Deng, K.Z.; Fan, H.D.; Du, S. Monitoring and Analysis of Surface Deformation in Mining Area Based on InSAR and GRACE. *Remote Sens.* **2018**, *10*, 1392. [CrossRef]
3. Benzaazoua, M.; Fall, M.; Belem, T. A contribution to understanding the hardening process of cemented pastefill. *Miner. Eng.* **2004**, *17*, 141–152. [CrossRef]
4. Hassani, F.; Archibald, J. *Mine Backfill*; Canadian Institute of Mining, Metallurgy and Petroleum: Montreal, QC, Canada, 1998.
5. Fall, M.; Celestin, J.C.; Pokharel, M.; Toure, M. A contribution to understanding the effects of curing temperature on the mechanical properties of mine cemented tailings backfill. *Eng. Geol.* **2010**, *114*, 397–413. [CrossRef]
6. Qi, C.; Fourie, A. Cemented paste backfill for mineral tailings management: Review and future perspectives. *Miner. Eng.* **2019**, *144*, 106025. [CrossRef]
7. Guo, S.J.; Zhang, J.X.; Li, M.; Zhou, N.; Song, W.J.; Wang, Z.J.; Qi, S.M. A preliminary study of solid-waste coal gangue based biomineralization as eco-friendly underground backfill material: Material preparation and macro-micro analyses. *Sci. Total Environ.* **2021**, *770*, 145241. [CrossRef] [PubMed]
8. Qi, C.-C. Big data management in the mining industry. *Int. J. Miner. Metall. Mater.* **2020**, *27*, 131–139. [CrossRef]
9. Tang, F.Q.; Lu, J.X.; Li, P.F. A prediction model for mining subsidence in loess-covered mountainous areas of western China. *Curr. Sci.* **2019**, *116*, 2036–2043. [CrossRef]
10. Shi, M.Y.; Yang, H.L.; Wang, B.C.; Peng, J.H.; Gao, Z.Z.; Zhang, B. Improving Boundary Constraint of Probability Integral Method in SBAS-InSAR for Deformation Monitoring in Mining Areas. *Remote Sens.* **2021**, *13*, 1497. [CrossRef]
11. Guo, G.L.; Zhu, X.J.; Zha, J.F.; Wang, Q. Subsidence prediction method based on equivalent mining height theory for solid backfilling mining. *Trans. Nonferrous Met. Soc. China* **2014**, *24*, 3302–3308. [CrossRef]
12. Guo, K.K.; Guo, G.L.; Li, H.Z.; Wang, C.Y.; Gong, Y.Q. Strata movement and surface subsidence prediction model of deep backfilling mining. *Energy Sources Part A Recover. Util. Environ. Eff.* **2020**, 1–15. [CrossRef]
13. Zhu, X.J.; Guo, G.L.; Liu, H.; Yang, X.Y. Surface subsidence prediction method of backfill-strip mining in coal mining. *Bull. Eng. Geol. Environ.* **2019**, *78*, 6235–6248. [CrossRef]
14. Vyazmensky, A.; Elmo, D.; Stead, D.; Rance, J.R. Combined finite-discrete element modelling of surface subsidence associated with block caving mining. In Proceedings of the 1st Canada—U.S. Rock Mechanics Symposium, Vancouver, BC, Canada, 27–31 May 2007; pp. 467–475.
15. Hamdi, P.; Stead, D.; Elmo, D.; Toyra, J. Use of an integrated finite/discrete element method-discrete fracture network approach to characterize surface subsidence associated with sub-level caving. *Int. J. Rock Mech. Min. Sci.* **2018**, *103*, 55–67. [CrossRef]
16. Li, Y.; Liu, L.M.; Li, Y.H.; Zhang, M.Y.; Wang, X.J. Quantitative Calculation Method Surface Subsidence Caused by Underground Mining Based on GIS. *Geotech. Geol. Eng.* **2020**, *38*, 61–69. [CrossRef]
17. Shi, H.; Zhang, Y.B.; Tang, L. Physical test of fracture development in the overburden strata above the goaf and diffusion process of permeable grout slurry. *Bull. Eng. Geol. Environ.* **2021**, *80*, 4791–4802. [CrossRef]

18. Lamich, D.; Marschalko, M.; Yilmaz, I.; Bednarova, P.; Niemiec, D.; Mikulenka, V.; Kubecka, K. Geodetic monitoring of roads as a tool for determination of hazard zones in areas influenced by deep coal mining. *Bull. Eng. Geol. Environ.* **2016**, *75*, 1033–1044. [CrossRef]
19. Ju, Y.; Nie, X.D.; Zhu, Y.; Xie, H.P. In situ fluidized mining and conversion solution to alleviate geological damage and greenhouse gas emissions due to coal exploitation: A numerical analysis and evaluation. *Energy Sci. Eng.* **2021**, *9*, 40–57. [CrossRef]
20. Huang, Q.X.; Cao, J. Research on Coal Pillar Malposition Distance Based on Coupling Control of Three-Field in Shallow Buried Closely Spaced Multi-Seam Mining, China. *Energies* **2019**, *12*, 462. [CrossRef]
21. Hu, H.F.; Lian, X.G.; Chen, S.Y. Effects on Buildings of Surface Curvature Caused by Underground Coal Mining. *J. Eng. Technol. Sci.* **2016**, *48*, 254–275. [CrossRef]
22. Guo, W.Y.; Hu, B.N.; Zhang, J.Y.; Chen, Q.T.; Jia, X.G.; Li, J.H. Research on the Settlement Regulation and Stability of Large Building Foundation over Gobs: A Case Study in the Xiangcheng Coal Mine, China. *Shock Vib.* **2021**, *2021*, 1–17.
23. Guo, W.B.; Guo, M.J.; Tan, Y.; Bai, E.H.; Zhao, G.B. Sustainable Development of Resources and the Environment: Mining-Induced Eco-Geological Environmental Damage and Mitigation Measures-A Case Study in the Henan Coal Mining Area, China. *Sustainability* **2019**, *11*, 4366. [CrossRef]
24. Liu, S.G.; Bai, J.B.; Wang, X.Y.; Wu, B.W.; Wu, W.D. Mechanisms of Floor Heave in Roadways Adjacent to a Goaf Caused by the Fracturing of a Competent Roof and Controlling Technology. *Shock Vib.* **2020**, *2020*, 1–17. [CrossRef]
25. Dong, J.H.; Li, H.J.; Wang, Y.S.; Zhang, Y.H. Characteristics and monitoring-based analysis on deformation mechanism of Jianshanying landslide, Guizhou Province, southwestern China. *Arab. J. Geosci.* **2021**, *14*, 1–10. [CrossRef]
26. Chen, D.D.; Wang, E.; Xie, S.R.; He, F.L.; Wang, L.; Zhang, Q.; Wu, X.Y.; Jiang, Z.S.; Li, Y.B.; Shi, S.H. Roadway Surrounding Rock under Multi-Coal-Seam Mining: Deviatoric Stress Evolution and Control Technology. *Adv. Civ. Eng.* **2020**, *2020*, 1–18.
27. Cheben, V.; Drusa, M.; Kuba, M. Innovative Groundwater Table Monitoring Using TDR Technology. *Int. J. Geomate* **2015**, *9*, 1428–1433.
28. Petlovanyi, M.V.; Ruskykh, V.V.; Zubko, S.A. Peculiarities of the underground mining of high-grade iron ores in anomalous geological conditions. *J. Geol. Geogr. Geoecol.* **2019**, *28*, 706–716. [CrossRef]
29. Marschalko, M.; Yilmaz, I.; Kristkova, V.; Fuka, M.; Kubecka, K.; Bouchal, T. An indicative method for determination of the most hazardous changes in slopes of the subsidence basins in underground coal mining area in Ostrava (Czech Republic). *Environ. Monit. Assess.* **2013**, *185*, 509–522. [CrossRef] [PubMed]
30. Dong, Y.; Huang, Y.C.; Du, J.F.; Zhao, F. Study on Overburden Stability and Development Height of Water Flowing Fractured Zone in Roadway Mining with Cemented Backfill. *Shock Vib.* **2021**, *2021*, 1–12.
31. Li, H.J.; Li, J.H.; Li, L.; Xu, H.; Wei, J.J. Prevention of water and sand inrush during mining of extremely thick coal seams under unconsolidated Cenozoic alluvium. *Bull. Eng. Geol. Environ.* **2020**, *79*, 3271–3283. [CrossRef]
32. Liu, W.R. Experimental and Numerical Study of Rock Stratum Movement Characteristics in Longwall Mining. *Shock Vib.* **2019**, *2019*, 1–15. [CrossRef]
33. Liu, S.G.; Bai, J.B.; Wang, X.Y.; Wu, B.W.; Wang, G.H.; Li, Y.H.; Xu, J. Study on Dynamic Evolution of Roof Crack and Support Timing of Secondary Tunneling for Large Section Open-Off Cut in Deep Mines. *Adv. Civ. Eng.* **2021**, *2021*, 1–16.
34. Akcin, H. A GIS-based building risk assessment for the subsidence due to undercity coal mining activities in Zonguldak, Turkey. *Arab. J. Geosci.* **2021**, *14*, 1–13. [CrossRef]
35. Kong, D.Z.; Li, Q.; Wang, N.; Wu, G.Y. Analysis on the Breaking Law of Soft and Thick Roof of Fully Mechanized Top-Coal Caving Face. *Geotech. Geol. Eng.* **2020**, *38*, 5941–5953. [CrossRef]
36. Ju, J.F.; Xu, J.L. Structural characteristics of key strata and strata behaviour of a fully mechanized longwall face with 7.0 m height chocks. *Int. J. Rock Mech. Min. Sci.* **2013**, *58*, 46–54. [CrossRef]
37. Guo, J.; Feng, G.R.; Wang, P.F.; Qi, T.Y.; Zhang, X.R.; Yan, Y.G. Roof Strata Behavior and Support Resistance Determination for Ultra-Thick Longwall Top Coal Caving Panel: A Case Study of the Tashan Coal Mine. *Energies* **2018**, *11*, 1041. [CrossRef]
38. Zhu, C.; Yuan, Y.; Yuan, C.F.; Liu, F.Q.; Chen, Z.S.; Wang, S.Z. Study on the structural forms of the key strata in the overburden of a stope during periodic weighting and the reasonable working resistance of the support. *Energy Sci. Eng.* **2020**, *8*, 2599–2620. [CrossRef]
39. ABAQUS. *Abaqus 2017 User's Manual*; SIMULIA: Providence, RI, USA, 2017.
40. Xu, Z. *A Concise Course of Elasticity*, 5th ed.; Higher Education Press: Beijing, China, 2016.
41. Liu, S.; Bai, J.; Wang, X.; Yan, S.; Zhao, J. Field and Numerical Study on Deformation and Failure Characteristics of Deep High-Stress Main Roadway in Dongpang Coal Mine. *Sustainability* **2021**, *13*, 8507. [CrossRef]
42. Shen, W.L.; Wang, M.; Cao, Z.Z.; Su, F.Q.; Nan, H.; Li, X.L. Mining-Induced Failure Criteria of Interactional Hard Roof Structures: A Case Study. *Energies* **2019**, *12*, 3016. [CrossRef]
43. *GB50936-2014, Technical Code for Concrete Filled Steel Tubular Structures*; China Architecture & Building Press: Beijing, China, 2014.
44. *GB50017-2017, Standard for Design of Steel Structures*; China Architecture & Building: Beijing, China, 2017.
45. Wang, Y.B.; Yin, D.W.; Chen, S.J.; Zhang, L.B.; Liu, D.Y.; Sun, Y.H. Experimental Study on Properties of Rock-Cemented Coal Gangue-Fly Ash Backfill Bimaterials with Different Coal Gangue Particle Sizes. *Adv. Civ. Eng.* **2020**, *2020*, 1–12. [CrossRef]
46. Xu, W.; Zhang, Y.; Zuo, X.; Hong, M. Time-dependent rheological and mechanical properties of silica fume modified cemented tailings backfill in low temperature environment. *Cem. Concr. Compos.* **2020**, *114*, 103804. [CrossRef]

47. Li, H.; Zhang, Y.; Xu, L.; Jia, X.; Gu, X. Examination of the treatment quality of filling mined-out voids using super-high-water material by the TEM technique. *Environ. Earth Sci.* **2017**, *76*, 115. [CrossRef]
48. Zhang, Z.Z.; Deng, M.; Bai, J.B.; Yan, S.; Yu, X.Y. Stability control of gob-side entry retained under the gob with close distance coal seams. *Int. J. Min. Sci. Technol.* **2021**, *31*, 321–332. [CrossRef]
49. Li, H.; Zhang, B.Y.; Bai, H.B.; Wu, J.J.; Meng, Q.B.; Xiao, N.; Li, F.; Wu, G.M. Surface Water Resource Protection in a Mining Process under Varying Strata ThicknessA Case Study of Buliangou Coal Mine, China. *Sustainability* **2018**, *10*, 4634. [CrossRef]
50. Malinowska, A.; Hejmanowski, R.; Dai, H.Y. Ground movements modeling applying adjusted influence function. *Int. J. Min. Sci. Technol.* **2020**, *30*, 243–249. [CrossRef]
51. Yan, W.T.; Dai, H.Y.; Chen, J.J. Surface crack and sand inrush disaster induced by high-strength mining: Example from the Shendong coal field, China. *Geosci. J.* **2018**, *22*, 347–357. [CrossRef]
52. Lu, G.D.; Yang, X.G.; Qi, S.C.; Li, X.L.; Ding, P.P.; Zhou, J.W. A generic framework for overpressure generation in sedimentary sequences under thermal perturbations. *Comput. Geotech.* **2020**, *124*, 103636. [CrossRef]
53. Liu, X.; Guo, G.; Li, H. Study on Damage of Shallow Foundation Building caused by Surface Curvature Deformation in Coal Mining Area. *KSCE J. Civ. Eng.* **2019**, *23*, 4601–4610. [CrossRef]
54. Li, L.; Aubertin, M.; Belem, T. Formulation of a three dimensional analytical solution to evaluate stresses in backfilled vertical narrow openings. *Can. Geotech. J.* **2005**, *42*, 1705–1717. [CrossRef]
55. Zhu, X.J.; Guo, G.L.; Wang, J.; Fang, Q.; Chen, T. Analysis of strata and ground subsidence in fully mechanized solid backfilling mining: A case study of Huayuan coal mine. *Trans. Inst. Min. Metall. Sect. A Min. Technol.* **2016**, *125*, 233–241. [CrossRef]
56. Zhang, X.S.; Yu, H.C.; Dong, J.Y.; Liu, S.P.; Huang, Z.Q.; Wang, J.X.; Wong, H. A physical and numerical model-based research on the subsidence features of overlying strata caused by coal mining in Henan, China. *Environ. Earth Sci.* **2017**, *76*, 1–11. [CrossRef]
57. Chen, Q.; Sun, S.; Liu, Y.; Qi, C.; Zhou, H.; Zhang, Q. Immobilization and leaching characteristics of fluoride from phosphogypsum-based cemented paste backfill. *Int. J. Miner. Metall. Mater.* **2021**, 1–13. [CrossRef]
58. Zhang, F.T.; Wang, X.Y.; Bai, J.B.; Wang, G.Y.; Wu, B.W. Post-peak mechanical characteristics of the high-water material for backfilling the gob-side entry retaining: From experiment to field application. *Arab. J. Geosci.* **2020**, *13*, 1–13. [CrossRef]
59. Zhang, J.X.; Li, M.; Taheri, A.; Zhang, W.Q.; Wu, Z.Y.; Song, W.J. Properties and Application of Backfill Materials in Coal Mines in China. *Minerals* **2019**, *9*, 53. [CrossRef]
60. Zhang, Z.Z.; Bai, J.B.; Han, Z.T.; Wang, X.Y.; Xu, Y.; Wang, M. Roof mechanics analysis and backfill technology for abandoned roadway. *J. Min. Saf. Eng.* **2013**, *30*, 194–198.
61. Wang, E.; Xie, S.R.; Chen, D.D.; Sun, Y.H.; Qi, P.Y. Influence of different loading rates on the mechanics and failure characteristics of high-water-content material. *Geomech. Geoengin.* **2021**, 1–12. [CrossRef]
62. Li, E.; Zhou, J.; Shi, X.; Jahed Armaghani, D.; Yu, Z.; Chen, X.; Huang, P. Developing a hybrid model of salp swarm algorithm-based support vector machine to predict the strength of fiber-reinforced cemented paste backfill. *Eng. Comput.* **2020**, 1–22. [CrossRef]
63. Wu, J.Y.; Jing, H.W.; Meng, Q.B.; Yin, Q.; Yu, L.Y. Assessment of cemented waste rock backfill for recycling gangue and controlling strata: Creep experiments and models. *Environ. Sci. Pollut. Res.* **2021**, *28*, 35924–35940. [CrossRef] [PubMed]

MDPI AG
Grosspeteranlage 5
4052 Basel
Switzerland
Tel.: +41 61 683 77 34

Minerals Editorial Office
E-mail: minerals@mdpi.com
www.mdpi.com/journal/minerals

www.ingramcontent.com/pod-product-compliance
Lightning Source LLC
Chambersburg PA
CBHW061619210926
51850CB00009B/2371
9783725823574